Kolla Bhanu Prakash
Editor

Harnessing AI for Drug Discovery

Machine Learning Perspectives

nova
science publishers
www.novapublishers.com

DOI: https://doi.org/10.52305/XIKL2761

Library of Congress Cataloging-in-Publication Data

ISBN: 979-8-89530-540-9 (Hardcover)
ISBN: 979-8-89530-664-2 (eBook)

Published by Nova Science Publishers, Inc. † New York

To my parents, Sri. Kolla Narayana Rao
and Smt. Kolla Uma Maheswari,
and my wife, Mrs. M. V. Prasanna Lakshmi

Contents

Preface

Developing new drugs is one of modern biomedicine's most complex and multi-step tasks. This process requires synthesizing multidisciplinary knowledge, ranging from molecular biology and biochemistry to clinical pharmacology and bioinformatics. Today, in the context of increasing disease complexity and the emergence of new therapeutic targets, traditional drug discovery methods are complemented by advanced technologies such as high-throughput screening (HTS), machine learning, and molecular modelling.

The main goal of this book is to provide readers with a deep understanding of all stages of the drug discovery process. We will consider the ways of identifying potential therapeutic targets and compounds, including methods of structural biology and rational drug design. Particular attention is paid to the process of optimizing lead compounds, which includes drug chemistry, toxicology, and pharmacokinetics, as well as adsorption, distribution, metabolism, and excretion (ADME) issues.

In recent decades, methods aimed at personalized medicine and targeted therapy have significantly developed. The book discusses cutting-edge approaches such as CRISPR genome editing, monoclonal antibody therapy, and the use of artificial intelligence to predict the efficacy and safety of new compounds. We also look at the principles of repositioning existing drugs and repurposing them for new indications.

An essential aspect of this book is the regulatory environment associated with developing and launching new drugs. We discuss the clinical trial process from Phase I to Phase III, including safety, efficacy, and post-marketing surveillance. We also consider the regulatory requirements and challenges facing pharmaceutical companies in different countries.

This book is intended for students, scientists, pharmacists, and anyone interested in drug discovery. We express our deep gratitude to all who contributed to its creation and to those who have dedicated their careers to the study, development, and implementation of new therapeutic methods. We hope this publication will be a valuable resource for all who seek to understand and participate in developing drugs that can change the future of medicine.

Dr. Kolla Bhanu Prakash

Acknowledgments

I thank the Almighty and my parents for their endless support, guidance, and love throughout all stages. I dedicate this book to my parents, family members, and my wife.

I would like to thank Sri. Koneru Satyanarayana, President of K. L. University, India, for his continuous support and encouragement throughout the preparation of this book.

I would like to express my gratitude to all who supported, shared, talked things over, read, wrote, offered comments, allowed us to quote remarks, and assisted in editing, proofreading, and designing throughout this book journey. We pay our sincere thanks to the open data set providers.

I am grateful to the Nova Publishing team, who showed us the ropes to creating this book. Without that knowledge, I would not have ventured into starting this book, which ultimately led to this. Their trust in us, their guidance, and their provision of the necessary time and resources gave us the freedom to manage this book.

Last, I'd like to thank my readers, who trusted us, and I hope this work inspires and guides them.

Dr. Kolla Bhanu Prakash

Chapter 1

Machine Learning and Drug Discovery: Supervised Learning Paradigms

Vidya Niranjan[1,*]
C. Lavanya[1]
Deepthika Kanagarajan[2]
Srirama P. Bhat[3]
and Sunidhi[1]
[1]Department of Biotechnology, RV College of Engineering,
affiliated to Visvesvaraya Technological University, Belagavi, India
[2]Department of Agriculture and Food Engineering, IIT Kharagpur, West Bengal, India
[3]Department of Biotechnology and Biochemical Engineering, IIT Kharagpur,
West Bengal, India

Abstract

The traditional drug discovery process is both costly and time-consuming, often requiring over a decade and billions of dollars to bring a new drug to market. here are stages in the process of drug discovery including identifying lead compounds, identifying targets, preclinical testing and clinical trials. Because of toxicity or an inability to be effective, many promising drugs fail at this stage. One potential approach to dealing with these challenges is through supervised learning techniques. These algorithms allow them to find patterns and make guesses using labelled data that can hasten drug development reduce costs as well as increase success rates of new drug candidates. At any particular stage of the drug discovery, supervised learning can be

* Corresponding Author's Email: vidya.n@rvce.edu.in

In: Harnessing AI for Drug Discovery
Editor: Kolla Bhanu Prakash
ISBN: 979-8-89530-540-9

employed. For example, when determining targets, these algorithms may utilize biological data to reveal proteins or genes associated with diseases. They also have the ability to identify big chemical libraries for picking lead compounds with certain activities biologically speaking. Also, machine-learning models based on supervised learning can foretell how a compound would behave if it were developed as a medicine and whether it would be toxic along its pathway helping select which ones go forward for further study. Supervised learning is not only efficient but also can cause a revolution in drug discovery and development.

Keywords: drug discovery, supervised learning, linear regression, logistic regression, neural networks, ensemble methods

Supervised Learning Algorithms in Drug Discovery

The traditional drug discovery process is expensive and lengthy, often undertaking billions of dollars and taking over a decade. The strategy involves stages like lead compound discovery, target identification, preclinical testing and clinical trials. However, conventional methods have a high failure rate, meaning potential drugs being abandoned in clinical trials due to unexpected toxicity and lack of efficacy (DiMasi et al., 2016). Consequently, there is a demand for more efficient and cost-effective drug development methods.

Supervised learning is an area of machine learning that offers a promising solution to the challenges in drug discovery. Using the labelled data to identify patterns and make predictions, these algorithms can remarkably accelerate the drug discovery process, reduce costs, and increase the success rate of new drug candidates (LeCun et al., 2015).

Supervised learning has a wide range of applications in drug discovery, including several key areas such as:

- *Target Identification:* Potential therapeutic targets such as disease-related proteins or genes, can be analysed using biological data using this algorithm (Ekins et al., 2014).
- *Lead Compound Selection:* An algorithm is capable of screening vast chemical libraries to identify lead compounds that show the desired biological activity (Goh et al., 2017)
- *Activity/Toxicity Prediction:* Supervised learning models can accurately predict a compound's biological activity and toxicity,

helping prioritize candidates for further investigation (Goodfellow et al., 2016).

Supervised learning improves the capacity to create safer and more effective medications while also streamlining the drug discovery process through the integration of these applications.

Linear and Logistic Regression

Linear Regression

Linear regression is a statistical technique that models the relationship between a dependent variable (y) and one or more independent variables (x). Its goal is to find the best-fitting straight line that represents this relationship, enabling predictions of the dependent variable based on the independent variables (James et al., 2013).

A straight-line equation is written as:

$$y = mx + b$$

In which:

"y is the dependent variable (the variable we want to predict)",
"x is the independent variable (the variable we use to make predictions)",
"m is the slope of the line (the rate at which y changes with respect to x)",
"b is the y-intercept (the value of y when x = 0)".

To find the values of "m" and "b" with minimum error between the actual and the predicted values (predicted by the line), we use linear regression. The most popular mean squared error is the cost function that is applied to this purpose by measuring the discrepancy between the values that were predicted and those that were actually obtained. MSE is defined as (James et al., 2013):

$$J(m, b) = \frac{1}{2n} \sum_{i=1}^{n} (y_i - (mx_i + b))^2$$

In which:

"n is the number of data points",

"y_i is the actual value of the dependent variable for data point i",
"x_i is the value of the independent variable for data point i",
"m and b are the slope and y-intercept of the line, respectively".

Linear regression aims to make the cost function as small as possible. It does this by using an optimization method called gradient descent. This process continuously adjusts the values of m and *b*, refining them to minimize the cost function until it reaches its lowest point.

Linear Regression in Drug Response Prediction

Linear regression plays a role in predicting drug response by linking a drug's molecular properties to its effects in the body. For instance, Li et al. (2018) used linear regression to predict a drug's binding affinity to target proteins based on its chemical structure, helping identify potential treatment candidates with high binding affinities.

Case Study: Predicting Drug Activity

Golbraikh and Tropsha (2002) conducted a significant study wherein linear regression to forecast the anti-HIV efficacy of several drugs was used. The effectiveness of quantitative structure-activity relationship (QSAR) models, especially when applied with linear regression approaches, in drug development is highlighted in this research.

Objective

To develop a QSAR model that predicts the anti-HIV activity of novel compounds and builds a mathematical connection between a compound's chemical properties and its biological activity against HIV. The methodology is mentioned in Figure 1.

Methodology

Results

Golbraikh and Tropsha's QSAR model predicted anti-HIV activity, with an R^2 value predicting strong correlation between the biological activity and molecular descriptors. The model's accuracy was depended on the robustness of linear regression and the proper selection of suitable molecular descriptors.

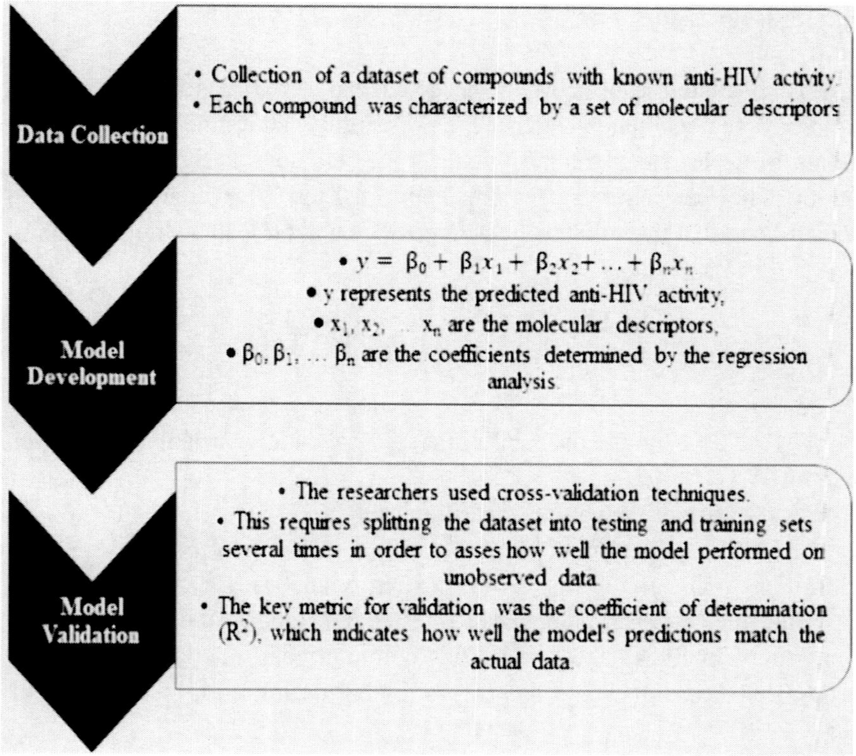

Figure 1. Methodology for the case study of linear regression in drug response prediction.

Conclusion

Golbraikh and Tropsha (2002) demonstrated the effectiveness of drug discovery using QSAR and linear regression models with the help of molecular descriptors to predict anti-HIV activity. The molecular models help to rapidly screen large drug libraries, by potentially speeding up the identification of potential therapeutic candidates and decreasing the cost of experimental testing and time. The study underlines the value of computational techniques in the latest drug discovery, with linear regression and well-selected molecular descriptors proving to be powerful tools for connecting chemical structure to biological activity, leading the way for therapeutic developments.

Logistic Regression

Logistic regression is used when the dependent variable is binary, such as yes/no or 0/1. It estimates whether the probability that a given specific input belongs to a particular category. The model employs a sigmoid (logistic) function, where the input value will be between 0 and 1, representing that there is a likelihood that the value belongs to positive class (Hosmer Jr et al., 2013):

$$P(y = 1 \mid x) = \frac{1}{1 + e^{-(mx+b)}}$$

In which:

"$P(y = 1 \mid x)$ is the probability that the dependent variable y is 1 given the input x",

"e is the base of the natural logarithm".

Similar to linear regression logistic regression uses a cost function to determine the difference between the actual labels and the expected probabilities. In logistic regression one common cost function is the cross-entropy loss function which has the following definition (Hosmer Jr et al. 2013)

$$J(m, b) = -\frac{1}{n} \sum_{i=1}^{n} [y_i \log(\hat{y}_i) + (1 - y_i) \log(1 - \hat{y}_i)]$$

In which:

"$\hat{y}_i$ is the predicted probability that y_i is 1",

"n is the number of data points",

"y_i is the actual label for data point i".

The goal of logistic regression is to minimize this cost function using gradient descent, similar to linear regression.

Logistic Regression in Drug Response Prediction

Drug response prediction also makes extensive use of logistic regression, particularly when the dependent variable is binary (drug response/non-response, for example). The probability of a drug producing a specific response based on its characteristics or attributes can be modelled using linear regression.

For instance, Zhang et al.'s study (2019) employed logistic regression to predict the outcomes of the cancer patient based on their genetic profiles. By evaluating the relationship between genetic markers and treatment response, the researchers identified the patients who had a benefit from specific treatments.

Case Study: Predicting Drug Toxicity

The study Chen et al. (2012) says that the logistic regression was used to predict drug hepatotoxicity. Analysis on chemical fingerprints and molecular characteristics gives the model that can accurately predict the hepatotoxicity of new drugs. The study pinpoints the effectiveness of logistic regression in assessing drug toxicity and safety.

Objectives

To develop model that can predict the potential for liver toxicity in new drugs and identify compounds that may cause liver damage early in the drug development process. The methodology is mentioned in Figure 2.

Methodology

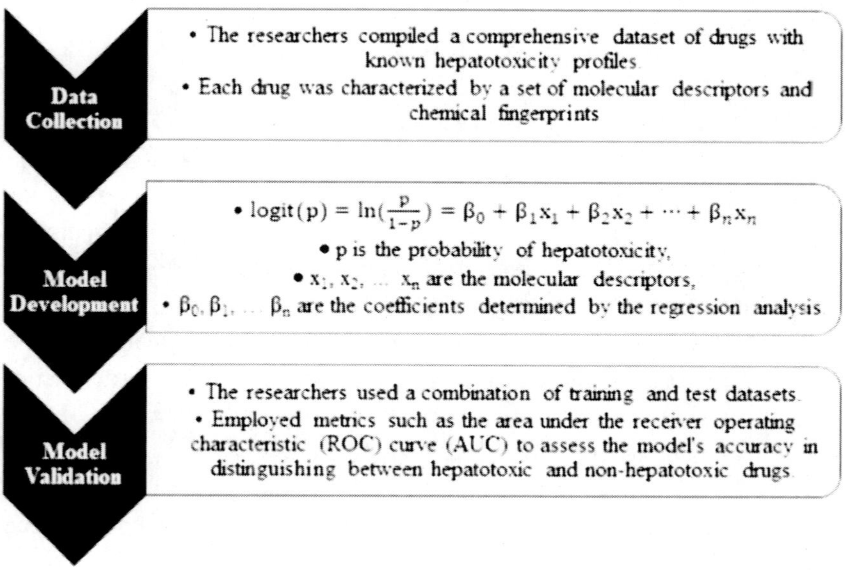

Figure 2. Methodology for case study of logistic regression in drug response prediction.

Results

The Chen et al., logistic regression model accurately predicted the hepatotoxicity in medicines, showing a strong ability to differentiate between the compounds that are hepatotoxic and non-hepatotoxic as indicated by a high AUC value. This verifies the effectiveness of both the logistic regression method and chosen chemical descriptors in predicting drug-induced liver damage.

Conclusion

Chen et al., (2012) demonstrated that logistic regression, using chemical fingerprints and molecular descriptors, effectively predicts hepatotoxicity of the medication. This pinpoints how computer models could enhance drug safety profiles during development.

This logistic regression model of this study is an essential tool for evaluating liver toxicity in medication candidates, lowering the risk of adverse outcomes in clinical trials and post-market surveillance. The strategy ensures safer drug releases and accelerates development. The study showcases the wide application of logistic regression in risk assessment and drug safety, enabling researchers to make better decisions and develop safer, more effective medicinal treatments through machine learning and statistical modelling.

Applications in Drug Response Prediction

- *Predicting Potency or Efficacy:* Both logistic and linear regression can be utilized to predict the potency or efficacy of treatment candidates by examining the relationship between pharmacological characteristics and biological activity.
- *Modelling Relationships:* These regression techniques help elucidate the complex connections between pharmacological characteristics and biological activity, enhancing the understanding of the underlying mechanisms of drug action.
- *Personalized Medicine:* Personalised medicines can be tailored according to the treatment of the patient by predicting the drug response based on individual trials.

Support Vector Machines (SVM)

Support vector machines (SVMs) are techniques for learning two-class discriminant functions from an array of training samples (Mammone et al., 2009). Introduced by Vapnik and colleagues in the 1990s, the SVMs are a supervised learning technique primarily used for predicting the property values in regression tasks (Heikamp & Bajorath, 2014a) and for binary data classification (Dara et al., 2022).

The method finds an optimal boundary in data space to separate the two classes by combining principles from convex optimization and statistical learning theory (Mammone et al., 2009). The fundamental principle of SVMs is the creation of a hyperplane that separates objects with distinct class labels, thereby deriving a separation "rule" for two classes of objects in high-dimensional descriptor reference spaces.

Despite their considerable potential for generalization, SVM models are often complex (Heikamp & Bajorath, 2014a). One crucial characteristic of SVMs is their attempt to use a technique known as structural risk minimization to minimize the error in training data and lower the computing complexity of models to prevent overfitting. To forecast new, unknown data, SVM models must strike a compromise between these two goals to perform well in generalization. SVMs have gained popularity recently for applications related to drug development, including property prediction, chemical classification and the hunt for new active compounds (Heikamp & Bajorath, 2014a).

Because SVM can rank compounds from each database, differentiate between active and inactive compounds, and train regression models, it is essential to the drug development process. Regression models use a query for datasets to predict, which is essential for understanding the link between the drug and ligand (Patel et al., 2020) (Burges, n.d.).

Mechanisms and Advantages in High-Dimensional Data Spaces

In both academia and industry statistical scientists frequently work with high-dimensional data. A feature or variable count may in some cases even surpass the total number of independent samples. Historically relatively few variables were used in the majority of statistical research. However, modern advances in computing power and data storage have ushered in the high-dimensional data revolution, now central to most statistical research (Narisetty, 2020).

High-dimensional data used in drug development is frequently derived from proteomic and genomic research, where each sample is characterized by hundreds or thousands of measurements typically collected simultaneously. This complexity arises from the interplay between biological systems, such as cancer biology, and the statistical and mathematical aspects of high-dimensional data spaces (Clarke et al., 2008).

There are many difficulties associated with analysing high-dimensional data in drug discovery. The 'big data' era has revealed challenges posed by the enormous number of variables gathered from a single sample. The "Curse of Dimensionality" (CoD) a concept introduced by Bellman, explains the challenges associated with high-dimensional data, including excessive space usage, increased computational effort, and difficulties in visualization (Altman & Krzywinski, 2018). Theoretically, more dimensions allow for more information storage, they often introduce more noise and redundancy (Venkat, 2018). High-dimensional datasets can lead to overfitting, where models capture noise rather than meaningful patterns, reducing their accuracy on new data. Identifying significant signals amidst noise is difficult due to potential duplication and irrelevance of many attributes (Vamathevan et al., 2019).

SVMs are well-suited for high-dimensional data due to their ability to handle numerous features and reveal complicated patterns. In contrast to traditional algorithms, SVMs use kernel functions, like the radial basis function (RBF), to transform data into higher-dimensional spaces where nonlinear relationships become linearly separable. This allows SVMs to identify intricate patterns that are not apparent in lower dimensions (Pavithra, 2017; Thudumu et al., 2020).

SVMs utilize support vectors—key data points closest to the decision boundary—to determine the optimal hyperplane that separates classes, resulting in efficient classification with low error rates (Cai et al., 2020). The kernel trick enables SVMs to perform complex transformations without explicitly calculating them, enhancing their ability to detect hidden patterns (Amini & Chang, n.d.). By maximizing the margin between classes, SVMs improve generalization, reduce overfitting, and are resilient to noise, making them effective for large, noisy datasets and real-time applications (Brereton & Lloyd, 2010; Maltarollo et al., 2019; Osuna et al., 1997).

SVMs are frequently used in the field of drug development for binary property or activity predictions, such as distinguishing compounds with or without specified activity, assessing synthetic accessibility and evaluating aqueous solubility. Initially, the compound libraries are mapped into a high-dimensional feature space, where molecules are represented as descriptor vectors and making them more likely to be linearly separable (Lavecchia, 2015).

SVM can be extrapolated in various ways to achieve objectives, with the main emphasis being on differentiating between molecules that are active and those that are not. The algorithm can be trained with several descriptors for feature selectors, such as target proteins and two-dimensional fingerprints, to control the process. Depending on which way a chemical is positioned relative to the hyperplane, a class label—either positive or negative—is created, classifying the compounds from most selective to least selective. Kernel functions plot the data in a higher-dimensional space where class separation is practical (Patel et al., 2020). Hyperplanes are implemented to separate samples as explained in the section above (a). in this context, distinct forms stand in for various classes of compounds, that can be differentiated based on their activity levels or classified as active or inactive. The Support Vector Machine linearly separates these classes by maximizing the distance between the closest samples of each class (known as support vectors) and the decision boundary (known as the hyperplane). Kernel functions, such as radial basis function (RBF), polynomial, linear, and sigmoid, can be used to project the original data into higher-dimensional space. The support vectors are positioned at the margin limits. Additionally, the distance between a sample (candidate compound) and the hyperplane can be used to calculate a ranking score. Choosing an appropriate kernel function is essential to ensure than an SVM performs well on data that is not linearly separable in its original dimensional space nut can be linearly separated in a higher-dimensional space. The number of features and cases in a dataset varies, different kernel functions produce different learning outcomes. However, according to Occams Razor, the simplest function – in this case, the linear kernel- should be tried first before exploring more complex equations. It is often recommended to select the kernel function and its parameters using either heuristic or exhaustive methods, as the choice depends on the complexity of the input data and there is ongoing debate about the effectiveness of specific kernel functions for each dataset (Maltarollo et al. (2019). The number of SVM-based drug discovery studies has significantly increased over the past decade, particularly those involving virtual screening and QSAR (Gertrudes et al. 2012). For instance, large training datasets were used to predict hERG and H1N1 neuraminidase inhibition, creating predictive models that prioritize new chemical entities foe experimental validation. The hERG training library is biased because it is derived from qHTS-positive findings, so the authors applied bootstrapping and jackknife resampling to enhace model quality (Sun et al. (2017). Alternatively, the conSSert software predicted secondary structures by combing the consensus of multiple SVM models. This consensus approach helped reduce the training bias introduced by different sets of descriptors in each individual model, thereby improving overall prediction accuracy (Kieslich et al. 2016).

Practical Examples in the Binary Classification of Compound Activity

Identifying active compounds for target proteins is crucial in the early stages of drug development. Recently, data-driven computational techniques have shown significant potential in chemical activity prediction (Tian et al., 2024). The SVM approach provides several methods for predicting compound activity. For example, binary class label predictions in SVM classification can be used to distinguish between active and inactive substances. Additionally, Support Vector Regression (SVR) can predict actual compound potency values, which are theoretically similar to traditional QSAR but differ methodologically (Heikamp & Bajorath, 2014b). Beyond target-specific applications, other research has focused on methodological challenges, particularly in benchmark computations on molecules active against multiple targets. For instance, Jorrisen and Gilson ranked database compounds using SVM classification scores, leading to the discovery of more chemically diverse molecules compared to fingerprint-based search techniques (Jorissen & Gilson, 2005). The cross-training with SVMs (ct-SVM) multi-label classification approach, introduced by Michielan et al., was used to predict drug selectivity and determine compound potency profiles against human adenosine receptor subtypes. In ct-SVM, binary classifiers were developed for drugs active against each specific target, and these classifiers were combined to produce a final label. Furthermore, to identify highly potent drugs, three distinct ct-SVM models were built at progressively lower potency levels and applies sequentially (Cheng et al., 2011).

Thai and Ecker delve into the effectiveness of binary QSAR models for predicting the drug activity, primarily focusing on hERG blockers, in different study. They provide valuable perceptions into how performance differs based on the type of descriptors used and the activities of compounds. By evaluating detailed assessment metrics, their findings help to refine QSAR modelling techniques for drug development and toxicity prediction. Understanding these complications allows stakeholders to select appropriate descriptors, optimize models, and apply suitable validation methods. This information is crucial for enhancing QSAR modelling approaches and improving the accuracy of chemical activity predictions in drug discovery and related fields (Thai & Ecker, 2008).

Gurulingappa et al. introduced a paradigm for concept-based binary classification drug class prediction. The technique worked well for categorizing medications by taking particular details out of scholarly publications. In drug classification scenarios, the concept-based approach proved to be better than structure-based methods. This method can be applied to identify secondary medication effects in system biology research (Gurulingappa et al., 2009).

Decision Trees and Ensemble Methods

Well-known machine learning (ML) models, decision trees (DTs), have been researched since the 1970s (Garey, 1972). DTs are often considered interpretable machine learning models (Freitas & Freitas, n.d.). This notion has led to significant efforts to develop DTs (and similar logic machine learning models) with attributes such as the number of nodes, maximum/average depth, etc., that are thought to be significant for interpretability (Aglin et al., 2020). Additionally, work has been done on utilizing (soft) decision trees to streamline or approximate sophisticated machine-learning models (Bastani et al., 2018).

However, new research indicates that interpretability should be correlated with DTs' level of shallowness. Decision tree-based techniques are particularly noteworthy among supervised learning approaches because they are non-parametric and combine interpretability, efficiency, and exceptional accuracy when applied to tree ensembles. Because the trained model outputs its information as a set of easily understood if-then rules, the decision tree can also be regarded as one of the most transparent learning methods (Nantasenamat et al., 2010).

Computational biology has made substantial use of decision trees (Geurts et al., 2009). In many scientific issues, data items must be categorized into one of a limited number of classes according to the properties of the data items. For instance, doctors use patient records, biopsies, and other tests to categorize tumours into various recognized cancer kinds.

Classifiers that predict class labels for data items are known as decision trees, and examples include C4.5, CART, and more recent versions. One advantage of decision trees is that it is inherently an important kind of classifier (Kingsford & Salzberg, 2008). Decision trees are typically constructed by examining a set of training samples for which the class labels are known. They are then used to label other, unseen, samples (Caruana & Niculescu-Mizil, 2006). Combined with ensemble approaches, decision trees can result in state-of-the-art performance for predictive accuracy, and can be trained to make extremely accurate predictions if they are trained on high quality data. Importantly, computationally, tree-based methods are also very scalable when applied to high-throughput data sets Kingsford & Salzberg, 2008).

Decision Trees: Interpretability and Decision-Making Process

The two main processes involved in developing a tree are growth and pruning. A decision tree has an edge where leaf nodes are located branching out to internal nodes from the root node (Nantasenamat et al. 2010). According to Salzberg (1994) decision trees are created incrementally with each new question node added after the labelled training examples. One straightforward question could accurately categorize a set of training examples into their respective groups. In the event that there isn't a question that achieves perfect separation we utilize one that can reasonably and clearly separate the cases (Kingsford and Salzberg 2008). In order to further divide the training items into smaller and smaller subgroups we recursively select questions to create a tree. Reducing the complexity of the learned trees is essential when using decision trees to prevent overfitting of the training set. If splitting reduces impurity by less than a negligible amount one of the simplest stopping criteria is to leave it alone. At the other extreme of mis-dimensionality we could simply construct the tree to its maximum height and continue growing it until either all of the leaves are pure or there are no more leaves to split. The challenge here is figuring out how to remove nodes from the tree in order to prune it and prevent it from overfitting the training set. If the classification error over a held-out set of training examples drops as a result this can be achieved by collapsing internal nodes into leaves. This model transparency and understandability to human experts are already some of the major reasons for their success in computational biology. To this respect, they are flexible, allowing to solve a large diversity of problems, and user-friendly, in the sense that even non-specialists can use them. That is, several variables helped the interpretability of decision trees. The graphical structure of a decision tree comes first.

Second, a decision tree usually only contains part of the attributes—instead of all of them—which helps users to focus their analysis on the most important ones. Besides, a hierarchical tree structure provides information on the relative merit of different qualities: in general, the lower in the tree any attribute, that is, farther away from the root an attribute is situated, the more relevant it is for classification. Note that an attribute may appear more than once in the same path from the root to a leaf. This certainly may be the case for several binary cuts on nodes that hold numeric attribute contents. In that case, depth of an attribute is then defined as the shallowest appearance of that attribute along that path. Now, if more than one occurrence of a certain attribute appeared in a path where sets of edges do not overlap, then an aggregate measure of the depths of the occurrences, such as the mean depth would need to be computed (Freitas & Freitas, n.d.). Since the input features determine the first decision, one starts by reading from the root node. From these features, each branch on the root node would then represent a condition. One should then follow the branches corresponding to conditions which the input data satisfies. Every node after it on this path reflects a portion of the data that satisfies the requirements specified by the nodes before it. The trip around the tree ends with a leaf node, which displays the anticipated result for the given input data. This result depends on the route followed through the tree and the circumstances at each node (Merkwirth et al., 2004).

Among their numerous advantages are their self-explanatory nature, ease of use, ability to translate into a set of rules, ability to handle both categorical and numeric variables, ability to handle data sets with errors or missing values, and nonparametric methodology (which makes no assumptions about the space distribution and the structure of the classifier). Nevertheless, decision trees also have drawbacks, such as the dependent variable's limited ability to have discrete values, its poor performance in the presence of numerous complex interactions, and its sensitivity to noise, irrelevant variables, and the training set (Maimon & Rokach, n.d.). Since only one decision trees can be employed, decision trees have limited predictive power in comparison with ensemble methods like random forests, which utilize multiple trees. In QSAR, modelling, decision tree ensembles are the most reliable, effective, and computationally efficient machine learning techniques (Nantasenamat et al., 2010).

Random Forests and Gradient Boosting Machines: Enhancing Prediction Accuracy and Robustness

High-throughput screening often yields highly unbalanced datasets, with few compounds showing biological activity and most showing none. This can lead classification algorithms to favour majority class. However, methods such as random forests, partial least squares (PLS), or stochastic gradient boosting can effectively handle the imbalance and resolve the obstacle (Nantasenamat et al., 2010).

It is also important to remember that while individual decision trees can be very effective classifiers, merging the output of several decision trees can frequently result in higher accuracy. Occasionally, decision tree ensembles rank among the best classifier types. There are two methods for integrating decision trees: boosting and random forests. In the random forests approach, a randomized tree-building algorithm grows a large number of distinct decision trees. To create a modified training set that is the same size as the original but contains some training elements, the training set is sampled with replacement (Kingsford & Salzberg, 2008). Another kind of ensemble-supervised machine-learning technique that can be applied to regression and classification issues is called gradient boosting (Schapire, 2003). Gradient boosting machines (GBM) have witnessed a sharp increase in popularity among decision tree ensembles in recent years, primarily due to their state-of-the-art performance in modelling tabular data and their outstanding performances in data science contests. In order to produce a high-performance ensemble, GBM iteratively aggregates predictive models so that each one offsets the errors from the preceding phase (Boldini et al., 2023). For this reason, complex cases are the main focus during training. The fact that a final model is formed based on numerous separate models is the primary reason ensemble techniques, including random forest, additional trees, gradient boosting, etc., are named such. One supervised learning algorithm that is adaptable, simple to use, and doesn't produce hyper-parameters is Random Forest. This method does classification reasonably well. A minimum number of trees must be constructed in Random Forest in order for all data to be classified within this range. Where each piece of data dramatically influences the amount (Usman Malik. 2018).

The Random Forest Algorithm using Scikit-Learn and Python describes that the minimal number of trees for each data set depends on the amount of breaker properties. The degree of accuracy is significantly impacted by the quantity of trees. Beginning with the bare minimum of trees, the accuracy achieved increases as the number of trees increases. The number of trees plus accuracy will remain constant, but there is a maximum accuracy limit that exists once accuracy is attained. The number of breaker attributes affects the accuracy that this algorithm generates. Low accuracy will be produced by the Random Forest, which utilizes the number of breaker attributes equal to the number of attributes available (Huljanah et al., 2019).

Gradient Based Machines (GBMs) employ a learning process that sequentially fits new models to improve the accuracy of the response variable estimate. The core concept behind his technique is to create new base learners that have the highest possible correlation with the ensemble's negative gradient of the loss function (Schapire, 2003). If the traditional squared- error loss function is used, the learning process focuses on fitting the successive errors; however, alternative loss functions can be applied to offer a different perspective. Researchers have a broad array of loss functions to choose from, and they can even design custom loss functions tailored to specific tasks. This adaptability allows GBMs to be customized for nearly any data-driven task, adding significant flexibility to model design. As a result, selecting the optimal loss function often involves a trial-and-error approach (Natekin & Knoll, 2013). Rusheng et al. discuss the two categories of neurorobotics applications for GBMs that show promise: ensemble-based neural simulations and high-accuracy pattern recognition applications. When evaluating pattern recognition problems, boosted temporal models such as HMM can be effectively used to assess tasks such as voice and motion recognition (Rusheng Hu et al., 2006).

Natekin and Knoll highlighted that the evaluation speed of GBMs is a concern due to high memory consumption. To make predictions with a fitted GBM model, each base learner in the ensemble must be evaluated. While each base learner is simple, obtaining quick predictions can be challenging when the ensemble is large. Therefore, practitioners may need to balance the number of function evaluations per time interval with model complexity when using GBMs in time-sensitive tasks. However, once the GBM ensemble is trained, parallelisation can be fully leveraged to speed up predictions (Natekin & Knoll, 2013).

Neural Networks

Neural networks get their name from neurons of the human brain. The neural network consists of neurons (or nodes) that are interconnected with each other and form a layer. These neurons process the data provided to them as examples and learn from it just as the human brain learns from its experience. The neural networks are trained by these examples and are later capable of performing complex tasks on their own based on the examples they have previously encountered. They are also known as Artificial Neural Networks (ANNs) and find themselves in various applications, including image and speech recognition, natural language processing, and bioinformatics, to name a few of their applications (Goodfellow et al., 2016).

Basic Structure of ANNs

Artificial Neural Network consists of three layers namely: "The input layer"; "Hidden layer"; "The output layer"

The neurons in the input layer represents features in the dataset and pass the data to hidden layers without any computations (Bishop, 2006). Hidden layers are located between the input and the output layers. The performance of the network is influenced by the number of neurons and the hidden layers present (Goodfellow et al., 2016). The output layer generates the network output and is the final layer (Bishop, 2006).

The primary role of the input layer is to receive data and pass it on to the hidden layer without any computations (Bishop, 2006). The hidden layer then take the data, process it, and apply weights and biases. These layers are crucial for detecting complex patterns and representations within the data (Goodfellow et al., 2016). Once the data is processed, the output layer produces the final result, with the number of neurons in this layer corresponding to the number of prediction classes or desired output dimensions (Bishop, 2006).

Activation Functions

Table 1. Activation Functions

Sigmoid	$\sigma(x) = \frac{1}{1+e^{-x}}$	Squashes input values to the range [0, 1]
ReLU (Rectified Linear Unit)	$f(x) = \max(0, x)$	Introduces sparsity and mitigates the vanishing gradient problem
tanh	$tanh(x) = \frac{e^{x} - e^{-x}}{e^{x} + e^{-x}}$	Squashes input values to the range [-1, 1]

(Nielsen, 2015).

The table above shows some of the activation functions that can be used to introduce non-linearity into the network for effective learning and modelling of complex data as mentioned in Table 1.

Backpropagation for Learning

Backpropagation is the primary algorithm used for training ANNs. It involves the following steps: The steps involved is shown in Figure 3.

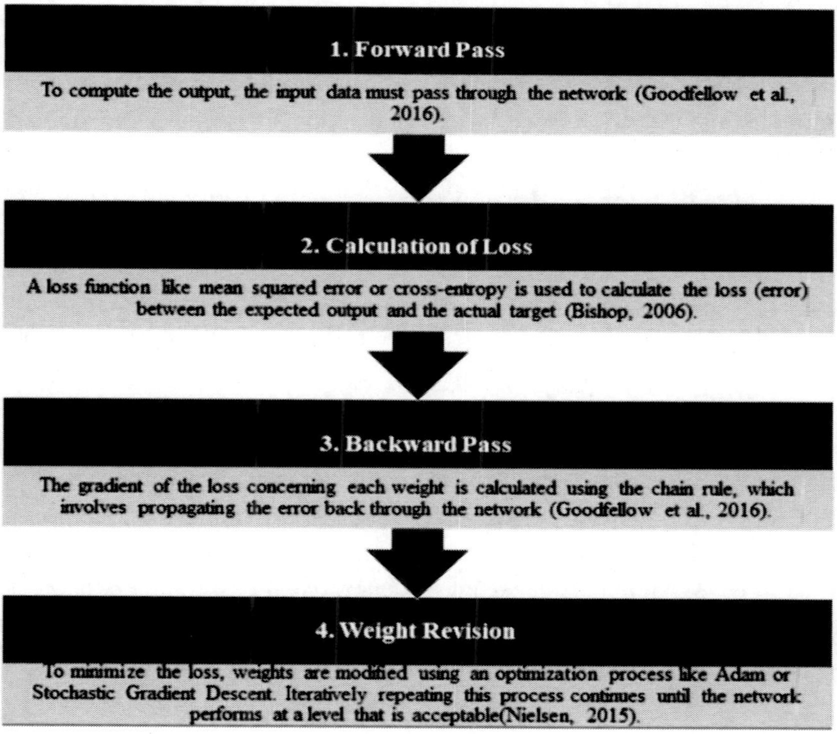

Figure 3. Steps involved in Backpropagation.

Advancements in Deep Learning

Over the past years, we have seen significant advancements in deep learning across various domains, especially in bioinformatics. Convolutional Neural

Networks (CNNs) and Recurrent Neural Networks (RNNs) are among the key neural networks that have immensely contributed to these breakthroughs.

Convolutional Neural Networks (CNNs)

Convolutional Neural Networks (CNNs) are specialised neural networks designed for analysing visual data, such as images, by extracting spatial hierarchies through convolutional operations. CNNs can identify local patterns like textures, edges and shapes, making them valuable for tasks such as image classification, object detection, and segmentation (Goodfellow et al. 2016). They are particularly effective in analysing protein structures and cellular images. For example, CNNs can predict the secondary structures of proteins by recognising patterns in amino acids sequences and comparing them to structural motifs (Zeng et al. 2016). By learning complex hierarchical representations, CNNs enhance the understanding of protein structures and functions.

Recurrent Neural Networks (RNNs)

Recurrent Neural Networks (RNNs) were developed with sequential data in mind. Traditional neural networks, such as feedforward networks, can only process the data that moves in a single direction from input to output. However, the RNNs are designed to form directed cycles and receive their previous output from previous time steps along with the current input for data processing. This feature makes them ideal for language modelling, speech recognition, and sequence prediction (Goodfellow et al., 2016).

Recurrent Neural Networks (RNNs) can used for bioinformatics purposes as well. It can be primarily used for sequence analysis purposes. RNNs can identify the regulatory elements in sequences such as promoter, operator, silencer, enhancer, etc and thereby predict the level of expression of a particular gene. They also help model the dynamics of biological processes (Lipton et al., 2015).

Applications of Deep Learning in Drug Discovery

As discussed earlier, deep learning has been found to be applicable in various fields, but its application in bioinformatics in the field of drug discovery has been tremendous over the past few years. Here are two fundamental applications:

Predicting Molecular Properties of Drug Candidates

In the initial stages of the drug development process, it is essential to identify the potential drug candidates with ideal molecular properties. Deep learning models can be used to predict the molecular properties of potential drug candidates like solubility, bioavailability, toxicity and binding affinity to target proteins. These models provide valuable insights into the molecular characteristics of a given drug candidate (Goodfellow et al., 2016).

Convolutional Neural Networks (CNNs) and Graph Neural Networks (GNNs) are well-suited for predicting the molecular properties of potential drug candidates, CNNs as previously mentioned, can evaluate visual data such as images or graphs, capturing spatial hierarchies that help in predicting molecular properties. GNNs, on the other hand, are adept at modeling the complex relationships between atoms and bonds within molecular structures (Rogers & Hahn, 2010). By processing large volumes of chemical data, these models achieve high accuracy in predicting molecular properties, thereby reducing the need for time-consuming and costly laboratory experiments (Goh et al., 2017).

De Novo Drug Design - Generating Novel Drug Structures

De novo drug design involves creating a drug component from the ground up, as the term suggests. This process utilizes deep learning models like Variational Autoencoders (VAEs) and Generative Adversarial Networks (GANs) to generate new chemical structures. These models ensure that the newly designed molecules possess essential properties, such as binding affinity, solubility, bioavailability, and other pharmacokinetic characteristics (Kingma & Welling, 2013).

Variational Autoencoders (VAEs) learn the latent representation of existing chemical structures and use this knowledge to generate novel compounds (Kingma & Welling, 2013). In contrast, Generative Adversarial Networks (GANs) create new molecules and then assess them, iteratively refining the quality of the generated compounds. GANs consist of two networks: a generator network that produces new molecules and a discriminator network that evaluates their quality (Kadurin et al., 2017).

Conclusion

Key Takeaways about Supervised Learning Algorithms and Their Impact on Drug Discovery

Supervised learning algorithms have entirely revolutionized the pharmaceutical industry and made it possible to predict and analyze complex biological and chemical data. Here are some of the takeaways from our discussion on different supervised learning algorithms:

- *Linear and Logistic Regression:* These form the basic models of predictive modeling since they yield results that we can comprehend and help to start models of greater complexity. In pharmacological effectiveness and toxicity studies, they are particularly useful in predicting binary outcomes (logistic regression) and continuous outcomes (linear regression) (Hosmer & Lemeshow, 2000).
- *Support Vector Machines (SVMs):* SVMs perform well in high-dimensional domains and particularly excel in classification problems. According to their chemical properties, they are useful in distinguishing two categories, e.g., active and inactive therapeutic candidates (Cortes & Vapnik, 1995).
- *Decision Trees and Ensemble Methods:* Decision trees are robust models used for modeling complex decision-making processes. Techniques like Random Forests and Gradient Boosting in ensembles enhance accuracy by effectively combining numerous models.

Drug candidates can be prioritized with the use of these techniques, which are very good at finding significant features and interactions in big datasets (Breiman, 2001; Friedman, 2001).

- *Neural Networks:* Neural networks, such as Recurrent Neural Networks (RNNs) and Convolutional Neural Networks (CNNs), have demonstrated remarkable ability in processing a wide range of data types, including gene sequences and chemical structures. They have proven to be extremely useful in predicting chemical characteristics and creating innovative drug structures due to their capacity to learn hierarchical and sequential patterns (Goodfellow et al., 2016).

Drug development processes are now much more precise and effective because of the incorporation of these algorithms, which also lowers costs and time spent bringing new drugs to the market.

Challenges and Future Directions

However, despite all these developments, there remain several challenges and future directions in the field of drug discovery using supervised learning algorithms:

Explainability: An important barrier is that many complex algorithms like neural networks are black box systems. To obtain trust from scientists and regulatory agencies, model interpretability needs to be enhanced. To resolve this issue, methods such as model-agnostic interpretability techniques or attention mechanisms are being investigated (Lipton, 2016).

Data Availability and Quality: Developing good models require well annotated high-quality datasets. Nonetheless, getting enough big data can be a challenge due to expensive experimental validation procedures and the proprietary nature of pharmaceutical research. Working together with public databases could manage this problem (LeCun et al., 2015).

Generalization: Another problem is ensuring that models work correctly on fresh untried data. Robust validation techniques combined with building models that can handle intricacies and variabilities

Integration with Experimental Methods: Computational predictions should be coupled with experimental validation. For discovering new drugs, supervised learning models can be made more reliable and useful by the use of iterative cycles of prediction and validation.

In future, one can expect application of reinforcement learning for optimizing drug development pipelines, using transfer learning and pre-training on related tasks, and developing complex models that could accommodate multi-omics data integration (AlQuraishi, 2021).

Acknowledgments

The authors would like to thank Akshay Uttarkar for his crucial insights in improving the presentation. Our heartfelt thanks to the staff and administration at R V College of Engineering for the support.

Author Contribution

Vidya Niranjan was in ideation, conceptualization and refining the manuscript. Lavanya C, Deepthika and Srirama contributed to the overall writing the manuscript. All the authors have read the final version of the manuscript and agreed for the publication

Conflict of Interest

The authors declare no conflict of interest.

Ethics Statement

No ethical consent is required.

References

Aghaei, S., Azizi, M. J., & Vayanos, P. (2019). Learning Optimal and Fair Decision Trees for Non-Discriminative Decision-Making. *Proceedings of the AAAI Conference on Artificial Intelligence,* 33(01), 1418–1426. https://doi.org/10.1609/aaai.v33i01.33011418.

Aglin, G., Nijssen, S., & Schaus, P. (2020). Learning Optimal Decision Trees Using Caching Branch-and-Bound Search. *Proceedings of the AAAI Conference on Artificial Intelligenc*e, 34(04), 3146–3153. https://doi.org/10.1609/aaai.v34i04.5711.

AlQuraishi, M. (2021). Machine learning in protein structure prediction. *Current Opinion in Chemical Biology*, 65, 1-8.

Altman, N., & Krzywinski, M. (2018). The curse(s) of dimensionality. Nature Methods, 15(6), 399–400. https://doi.org/10.1038/s41592-018-0019-x.

Amini, M., & Chang, S. (n.d.). *A review of machine learning approaches for high dimensional process monitoring.*

Bastani, O., Kim, C., & Bastani, H. (2018). Interpretability via Model Extraction (arXiv:1706.09773). *arXiv*. http://arxiv.org/abs/1706.09773.

Bishop, C. M. (2006). *Pattern Recognition and Machine Learning*. Springer.

Boldini, D., Grisoni, F., Kuhn, D., Friedrich, L., & Sieber, S. A. (2023). Practical guidelines for the use of gradient boosting for molecular property prediction. *Journal of Cheminformatics,* 15(1), 73. https://doi.org/10.1186/s13321-023-00743-7.

Breiman, L. (2001). *Random forests. Machine Learning*, 45(1), 5-32.

Brereton, R. G., & Lloyd, G. R. (2010). Support Vector Machines for classification and regression. *The Analyst,* 135(2), 230–267. https://doi.org/10.1039/B918972F.

Burges, C. J. C. (n.d.). A Tutorial on Support Vector Machines for Pattern Recognition.

Cai, C., Wang, S., Xu, Y., Zhang, W., Tang, K., Ouyang, Q., Lai, L., & Pei, J. (2020). Transfer Learning for Drug Discovery. *Journal of Medicinal Chemistry*, 63(16), 8683–8694. https://doi.org/10.1021/acs.jmedchem.9b02147.

Caruana, R., & Niculescu-Mizil, A. (2006). An empirical comparison of supervised learning algorithms. *Proceedings of the 23rd International Conference on Machine Learning - ICML* '06, 161–168. https://doi.org/10.1145/1143844.1143865.

Chen, M., Suzuki, A., Thakkar, S., Yu, K., Hu, C., Tong, W., DILIrank: The largest reference drug list ranked by the risk for developing drug-induced liver injury in humans, Drug Discov. *Today*, 2016, 21(4), 648-653.

Chen, T., Shang, C., Su, P., Keravnou-Papailiou, E., Zhao, Y., Antoniou, G., & Shen, Q. (2021). A Decision Tree-Initialised Neuro-fuzzy Approach for Clinical Decision Support. A*rtificial Intelligence in Medicine,* 111, 101986. https://doi.org/10.1016/j.artmed.2020.101986.

Cheng, T., Li, Q., Wang, Y., & Bryant, S. H. (2011). Binary Classification of Aqueous Solubility Using Support Vector Machines with Reduction and Recombination Feature Selection. *Journal of Chemical Information and Modeling*, 51(2), 229–236. https://doi.org/10.1021/ci100364a.

Clarke, R., Ressom, H. W., Wang, A., Xuan, J., Liu, M. C., Gehan, E. A., & Wang, Y. (2008). The properties of high-dimensional data spaces: Implications for exploring gene and protein expression data. *Nature Reviews Cancer,* 8(1), 37–49. https://doi.org/10.1038/nrc2294.

Cortes, C., & Vapnik, V. (1995). Support-vector networks. *Machine Learning*, 20(3), 273-297.

Dara, S., Dhamercherla, S., Jadav, S. S., Babu, C. M., & Ahsan, M. J. (2022). Machine Learning in Drug Discovery: A Review. *Artificial Intelligence Review*, 55(3), 1947–1999. https://doi.org/10.1007/s10462-021-10058-4.

DiMasi, J. A., Grabowski, H. G., & Hansen, R. W. (2016). Innovation in the pharmaceutical industry: New estimates of R&D costs. *Journal of Health Economics*, 47, 20-33.

Ekins, S., Mestres, J., & Testa, B. (2014). In silico pharmacology for drug discovery: methods for virtual ligand screening and profiling. *British Journal of Pharmacology,* 152(1), 9-20.

Elhai, R., Sherif, Y., Al-Mallah, M., & Sakr, S. (2021). Interpretability in healthcare: A comparative study of local machine learning interpretability techniques. *Computational Intelligence*, 37(4), 1633–1650. https://doi.org/10.1111/coin.12410.

Freitas, A. A., & Freitas, A. A. (n.d.). Comprehensible Classification Models – a position paper. 15(1).

Friedman, J. H. (2001). Greedy function approximation: A gradient boosting machine. *Annals of Statistics*, 29(5), 1189-1232.

García, S., Fernández, A., & Herrera, F. (2009). Enhancing the effectiveness and interpretability of decision tree and rule induction classifiers with evolutionary training set selection over imbalanced problems. *Applied Soft Computing*, 9(4), 1304–1314. https://doi.org/10.1016/j.asoc.2009.04.004.

Garey, M. R. (1972). Optimal Binary Identification Procedures. *SIAM Journal on Applied Mathematics,* 23(2), 173–186. https://doi.org/10.1137/0123019.

Gertrudes, J. C., Maltarollo, V. G., Silva, R. A., Oliveira, P. R., Honorio, K. M., & Da Silva, A. B. F. (2012). Machine Learning Techniques and Drug Design. *Current Medicinal Chemistry,* 19(25), 4289–4297. https://doi.org/10.2174/0929867 12802884259.

Geurts, P., Irrthum, A., & Wehenkel, L. (2009). Supervised learning with decision tree-based methods in computational and systems biology. *Molecular BioSystems*, 5(12), 1593. https://doi.org/10.1039/b907946g.

Goh, G. B., Hodas, N. O., & Vishnu, A. (2017). Deep learning for computational chemistry. *Journal of Computational Chemistry*, 38(16), 1291-1307.

Gonzalez-Cava, J. M., Reboso, J. A., Casteleiro-Roca, J. L., Calvo-Rolle, J. L., & Méndez Pérez, J. A. (2018). A Novel Fuzzy Algorithm to Introduce New Variables in the Drug Supply Decision-Making Process in Medicine. *Complexity,* 2018, 1–15. https://doi.org/10.1155/2018/9012720.

Golbraikh, A., Tropsha, A., Beware of q2!, *J. Mol. Graphics Modell.*, 2002, 20(4), 269-276.

Goodfellow, I., Bengio, Y., & Courville, A. (2016). *Deep Learning*. MIT Press.

Gurulingappa, H., Kolářik, C., Hofmann-Apitius, M., & Fluck, J. (2009). Concept-Based Semi-Automatic Classification of Drugs. *Journal of Chemical Information and Modelling*, 49(8), 1986–1992. https://doi.org/10.1021/ci9000844.

Hammann, F., Gutmann, H., Vogt, N., Helma, C., & Drewe, J. (2010). Prediction of Adverse Drug Reactions Using Decision Tree Modelling. *Clinical Pharmacology & Therapeutics,* 88(1), 52–59. https://doi.org/10.1038/clpt.2009.235.

Ho, T. K. (1995). Random Decision Forests. *Proceedings of 3rd International Conference on Document Analysis and Recognition*, 1, 278-282.

Hutter, F., Kotthoff, L., & Vanschoren, J. (Eds.). (2019). *Automated Machine Learning: Methods, Systems, Challenges*. Springer International Publishing. https://doi.org/ 10.1007/978-3-030-05318-5.

Jain, A. K., Duin, R. P. W., & Mao, J. (2000). Statistical pattern recognition: A review. *IEEE Transactions on Pattern Analysis and Machine Intelligence*, 22(1), 4-37.

Jolliffe, I. T. (2002). *Principal Component Analysis* (2nd ed.). Springer Series in Statistics. https://doi.org/10.1007/b98835.

Källblad, P., Zhao, Z., Wägberg, T., Hovmöller, M., Andersson, C., & Norinder, U. (2020). Interpretability and Predictive Modelling in Drug Discovery. *Molecular Informatics*, 39(1–2), 1900132. https://doi.org/10.1002/minf.201900132.

Ke, G., Meng, Q., Finley, T., Wang, T., Chen, W., Ma, W., ... & Liu, T. Y. (2017). LightGBM: A Highly Efficient Gradient Boosting Decision Tree. *Advances in Neural Information Processing Systems*, 30.

Kingma, D. P., & Ba, J. (2017). Adam: A Method for Stochastic Optimization (arXiv:1412.6980). *arXiv*. http://arxiv.org/abs/1412.6980.

Kuhn, M., & Johnson, K. (2013). Applied Predictive Modelling. Springer.

Le, T. T., & Huerta, G. (2019). A review of semi-supervised learning approaches in the era of deep learning. *Applied Intelligence,* 49(8), 2975–2988. https://doi.org/10.1007/ s10489-019-01531-x.

Lee, I. H., Choi, W., & Yoo, S. Y. (2014). Predicting drug toxicity using deep learning. *Current Pharmaceutical Design*, 20(7), 1028-1034.

LeCun, Y., Bengio, Y., & Hinton, G. (2015). *Deep Learning. Nature*, 521(7553), 436–444. https://doi.org/10.1038/nature14539.

Lee, M., Kim, D., Park, M., Jeong, M., & Choi, H. (2020). Prediction of Adverse Drug Reactions Using Comprehensive Modelling. *Clinical Pharmacology & Therapeutics*, 107(4), 953–962. https://doi.org/10.1002/cpt.1785.

Letham, B., Rudin, C., McCormick, T. H., & Madigan, D. (2015). Interpretable classifiers using rules and Bayesian analysis: Building a better stroke prediction model. *Annals of Applied Statistics*, 9(3), 1350–1371. https://doi.org/10.1214/15-AOAS848.

Li, Z., & He, L. (2017). A comprehensive review of definitions of interpretable machine learning and evaluation metrics. Proceedings of the 2020 *IEEE International Conference on Big Data* (Big Data), 285–294. .https://doi.org/10.1109/BigData50022.2020.9378133.

Li, Z., Yang, Y., Wei, M., & Lin, S. (2019). Interpretable drug adverse reaction prediction using knowledge graph embedding. *Journal of Biomedical Informatics*, 92, 103138. https://doi.org/10.1016/j.jbi.2019.103138.

Lipton, Z. C. (2018). *The Mythos of Model Interpretability* (arXiv:1606.03490). *arXiv*. http://arxiv.org/abs/1606.03490.

Lundberg, S. M., & Lee, S. I. (2017). A Unified Approach to Interpreting Model Predictions. *Advances in Neural Information Processing Systems*, 30.

Ma, X., & Sun, H. (2011). Support Vector Machine and Its Applications in Chemistry. *Chemometrics and Intelligent Laboratory Systems,* 110(2), 218–230. https://doi.org/10.1016/j.chemolab.2011.10.010.

Mattioni, B., Beck, B., Clevert, D. A., & Tyrchan, C. (2021). Addressing Interpretability in Drug Discovery with Attention-based Convolutional Neural Networks. *Journal of Cheminformatics,* 13(1), 31. https://doi.org/10.1186/s13321-021-00500-3.

Mehta, N., & Pandit, S. (2022). Computational Methods in Drug Discovery. In J. M. Walker (Ed.), *Computational Methods for Drug Discovery* (pp. 1–14). Springer. https://doi.org/10.1007/978-1-0716-2523-8_1.

Mehrjou, A., Hashimoto, T., Sato, I., & Sugiyama, M. (2018). Improved Interpretability of the Predictions of Multi-Task Deep Neural Networks (arXiv:1809.06362). *arXiv*. http://arxiv.org/abs/1809.06362.

Meredig, B., Agrawal, A., Kirklin, S., Saal, J. E., Doak, J. W., Thompson, A., Zhang, K., Choudhary, A., & Wolverton, C. (2014). Combinatorial screening for new materials in unconstrained composition space with machine learning. *Physical Review B*, 89(9), 094104.

Muegge, I., & Mukherjee, P. (2016). An overview of molecular fingerprint similarity search in virtual screening. *Expert Opinion on Drug Discovery*, 11(2), 137-148.

Murdoch, W. J., Singh, C., Kumbier, K., Abbasi-Asl, R., & Yu, B. (2019). Definitions, Methods, and Applications in Interpretable Machine Learning. *Proceedings of the National Academy of Sciences*, 116(44), 22071–22080. https://doi.org/10.1073/pnas.1900654116.

Pearl, J. (2009). *Causality: Models, Reasoning, and Inference* (2nd ed.). Cambridge University Press.

Polley, E. C., van der Laan, M. J., & Hubbard, A. E. (2011). Super Learner in Prediction. U.C. Berkeley Division of Biostatistics Working Paper Series. Working Paper 266.

Pu, L., Govindaraj, R. G., Lemoine, J. M., Wu, H. C., & Chen, Y. P. P. (2019). Deep Drug Target Learning. *Computers in Biology and Medicine*, 109, 170–177. https://doi.org/10.1016/j.compbiomed.2019.04.017.

Quinlan, J. R. (1986). Induction of decision trees. *Machine Learning*, 1(1), 81–106. https://doi.org/10.1007/BF00116251.

Rajkomar, A., Dean, J., & Kohane, I. (2019). Machine Learning in Medicine. *New England Journal of Medicine,* 380(14), 1347–1358. https://doi.org/10.1056/NEJMra1814259.

Rish, I. (2001). An Empirical Study of the Naive Bayes Classifier. *Proceedings of the IJCAI-01 Workshop on Empirical Methods in Artificial Intelligence*, 3.

Rupp, M. (2015). Machine learning for quantum mechanics in a nutshell. *International Journal of Quantum Chemistry*, 115(16), 1058-1073.

Segler, M. H. S., Kogej, T., Tyrchan, C., & Waller, M. P. (2018). Generating Focused Molecule Libraries for Drug Discovery with Recurrent Neural Networks. ACS Central *Science,* 4(1), 120–131. https://doi.org/10.1021/acscentsci.7b00403.

Sheridan, R. P., Wang, W. M., Cross, K. P., & Gedeck, P. (2013). Where might we find new activities? A self-organizing map and Bayesian analysis of 6.5 million activity measurements. *Journal of Computer-Aided Molecular Design*, 27(9), 755-768.

Shrikumar, A., Greenside, P., & Kundaje, A. (2017). Learning Important Features through Propagating Activation Differences (arXiv:1704.02685). *arXiv*. http://arxiv.org/abs/1704.02685.

Shrivastava, A., Pike, M. J., & Granger, E. (2014). Improving Classifier Diversity Using Regularized Linear Discriminant Analysis. *IEEE Transactions on Cybernetics*, 45(1), 115-126.

Stein, S. E., & Scott, D. R. (1994). Optimization and testing of mass spectral library search algorithms for compound identification. *Journal of the American Society for Mass Spectrometry,* 5(9), 859-866.

Swamidass, S. J., Chen, J., Bruand, J., Phung, P., Ralaivola, L., & Baldi, P. (2005). Kernels for Small Molecules and the Prediction of Mutagenicity, Toxicity and Anti-Cancer Activity. *Bioinformatics,* 21(1), 359–368. https://doi.org/10.1093/bioinformatics/bti615.

Tibshirani, R. (1996). Regression Shrinkage and Selection via the Lasso. *Journal of the Royal Statistical Society: Series B (Methodological),* 58(1), 267–288. https://doi.org/10.1111/j.2517-6161.1996.tb02080.x.

Van Der Walt, S., Colbert, S. C., & Varoquaux, G. (2011). The NumPy Array: A Structure for Efficient Numerical Computation. Computing in Science & Engineering, 13(2), 22-30. https://doi.org/10.1109/MCSE.2011.37.

Vaswani, A., Shazeer, N., Parmar, N., Uszkoreit, J., Jones, L., Gomez, A. N., Kaiser, Ł., & Polosukhin, I. (2017). Attention is All You Need (arXiv:1706.03762). *arXiv.* http://arxiv.org/abs/1706.03762.

Vaswani, A., & Polosukhin, I. (2018). The Transformer Model for NLP. In J. E. Hopcroft & D. B. Johnson (Eds.), *Advances in Neural Information Processing Systems* 31 (pp. 5998–6008). MIT Press. https://doi.org/10.7551/mitpress/11258.001.0001.

Venkat, N. (2018). *The Curse of Dimensionality: Inside Out. Inside Out.*

Vinyals, O., Fortunato, M., & Jaitly, N. (2015). Pointer Networks. *Advances in Neural Information Processing Systems*, 28.

Wang, M., Xue, Y., Zhao, Z., & Zhang, Y. (2019). Drug Adverse Reaction Prediction Based on Bayesian Network and Text Mining. *International Journal of Machine Learning and Computing*, 9(1), 93–99. https://doi.org/10.18178/ijmlc.2019.9.1.772.

Wang, Y., Yao, Q., Kwok, J. T., & Ni, L. M. (2020). Generalizing from a few examples: A survey on few-shot learning. *ACM Computing Surveys*, 53(3), 63:1-63:34. https://doi.org/10.1145/3386252.

Williams, R. J. (1992). Simple Statistical Gradient-Following Algorithms for Connectionist Reinforcement Learning. *Machine Learning*, 8, 229–256. https://doi.org/10.1007/BF00992696.

Zhou, Y., Li, W., Zhang, L., & Li, J. (2019). Improving Drug Adverse Reaction Prediction with Network Embedding and Text Mining. *Journal of Biomedical Informatics*, 98, 103291. https://doi.org/10.1016/j.jbi.2019.103291.

Chapter 2

Introduction to Machine Learning in Drug Discovery

B. Aarthirashmi*
B. Varsha
and M. Sri Vidya
Department of Bioscience, Sri Krishna Arts and Science College, Coimbatore, Tamil Nadu, India

Abstract

The incorporation of machine learning (ML) into drug discovery has revolutionized the pharmaceutical sector, presenting a novel approach to identifying effective therapeutic molecules. Traditionally, drug discovery involved a time-consuming, costly, and labor-intensive process marked by trial and error. Researchers synthesized numerous compounds and subjected them to various biological assays, often encountering high failure rates. However, ML introduces a paradigm shift, enabling researchers to navigate the vast chemical space more efficiently. ML systems are excellent at processing large datasets and identifying complex patterns that human researchers would miss. These algorithms evaluate vast amounts of chemical and biological data in drug discovery in order to precisely forecast the pharmacokinetic and pharmacodynamic characteristics of possible molecules. This predictive capability allows scientists to anticipate how drugs will behave in the body, identifying promising candidates earlier in the development process. One of the primary advantages of employing ML in drug discovery is the substantial reduction in time and costs. Researchers can

* Corresponding Author's Email: aarthirashmib@skasc.ac.in

In: Harnessing AI for Drug Discovery
Editor: Kolla Bhanu Prakash
ISBN: 979-8-89530-540-9

focus their efforts on potential leads by identifying likely successful compounds. This speeds up the discovery process and saves money that would have been used to investigate unsuccessful compounds. Additionally, ML optimizes drug development by offering insights into molecular structures conducive to successful therapies, departing from the traditional method of modifying existing molecules in hopes of efficacy enhancement. The applications of ML in drug discovery are extensive, ranging from identifying new drug targets to predicting adverse reactions and facilitating personalized medicine based on individual genetic profiles. In conclusion, the use of machine learning (ML) in drug discovery is a substantial breakthrough that improves effectiveness, accuracy, and affordability. This transformative approach holds promise for expediting the introduction of new therapies to the market, underscoring ML's potential to revolutionize pharmaceutical research and development.

Keywords: machine learning, drug discovery, pharmaceutical research, artificial intelligence, deep learning, reinforcement learning

Introduction

As a branch of artificial intelligence (AI), machine learning (ML) focuses on creating algorithms that let computers analyze, interpret, and learn from data. In contrast to traditional programming, which demands explicit instructions, machine learning algorithms automatically become more efficient over time. The following are some of the primary areas of machine learning: Supervised Learning: Labeled data is used to train models that forecast results for fresh data. Common algorithms include linear regression, decision trees, and neural networks. The technique of training models on unlabeled response data to reveal innate structures or hidden patterns is known as unsupervised learning. Typical examples include dimensionality reduction techniques like principal component analysis and clustering techniques like K-means. Reinforcement learning is a method for teaching agents how to make decisions byrewarding desired behavior and penalizing undesirable behavior. This is especially helpful in situations that call for strategic choices, like controlling robots or playing video games. Because machine learning can handle and analyze large amounts of data, it is instrumental in complex domains like drug discovery.

Traditional Drug Discovery Process

Traditionally, drug discovery is a lengthy and costly process, often taking over a decade and billions of dollars to bring a new drug to market. The steps include:

Target Identification

Target Identification and Validation Finding: Confirming possible targets is one of the most crucial initial phases in the drug discovery process. Machine learning, which finds potential drug targets by analyzing massive biological databases, significantly enhances this stage. Genomic Data Analysis: Machine learning algorithms can analyze enormous volumes of genomic data to identify genes connectedto specific disorders. Gene expression profiling and genome-wide association studies (GWAS) are two techniques that can be used to find new therapeutic targets. Network-Based Methodologies: By examining biological networks, machine learning might pinpoint crucial proteins or genes implicated in disease pathways. These networks include metabolic networks, signaling pathways, and networks involving interactions between proteins.

ML models highlight possible therapeutic targets by predicting which nodes (genes or proteins) are essential for the course of the disease. These approaches reduce the time and cost associated with target discovery and validation using traditional experimental methods.

Application

- *Data mining and Bioinformatics*: Machine learning algorithms can analyze biological data, including proteomics and genomes, to identify new treatment targets.
- *Predictive Modelling*: Machine learning models can forecast whether or not a given protein or gene willbe adequate.

Hit Discovery

Molecular Simulation and Medicinal Design: Once possible targets have been identified, the next step is todevelop compounds that can interact with them effectively. Machine learning aids with this by predicting the properties of new substances and generating creative treatment options.

Quantitative Structure-Action-relationships, or QSAR models, forecast a compound's biological action based on its chemical structure. Byrecognizing intricate connections between molecular characteristics and biological activity, machine learning improves QSAR models and produces more precise predictions. De Novo Pharmaceutical Design: ML methods, including highly generative models like generative adversarial networks (GANs) and variational autoencoders (VAEs), can produce new chemical structureswith desired features.

These models can generate a large number of potential drug candidates, and the most promising ones canbe selected for further investigation. Machine learning enables more efficient chemical space research, increasing the likelihood of finding new medication candidates.

Application

- *Virtual Screening:* Much less physical screening is needed when machine learning is used to forecast which chemicals are most likely to attach to the target. Quantitative Structure-Activity Relationships, or QSAR Models, predict a chemical compound's activity based on its structure.

Lead Optimization

Virtual Screening and Lead Optimization: Virtual screening is the technique of examining large chemical libraries to identify those that have the highest likelihood of binding to a target protein. Virtual screening becomes faster and more accurate with machine learning. Extensive Screening (HTS) Data analysis: By processing HTS data, machine learning models can accurately predict which compounds will bind to the target. This process dramatically reduces the number of compounds that require experimental testing. Docking Models: By forecasting the binding affinities of compounds to their targets, machine learning improves molecular docking. More sophisticated models can replicate the docking process with greater accuracy, locating lead compounds with strong binding affinities. These molecules can be further optimized by machine learning to improve their pharmacokinetic, selectivity, and effectiveness properties after lead identification, preparing them for clinical development.

Application

- *Pharmacokinetic and Toxicological Property Prediction*: Machine learning can predict the absorption, distribution, metabolism, and excretion (ADME) of a drug as well as its possible toxicity. Optimisation Algorithms: Machine learning (ML) can help create molecules with the desired attributes by predicting the outcomes of chemical changes.

Preclinical Testing

Predictive Modeling for ADMET and Clinical Trials: Just as crucial to a drug's success as its efficacy are its pharmacokinetic and safety characteristics. Early in the drug development process, machine learning is essential for predicting properties such as Absorption, Distribution, Metabolism, Excretion, and Toxicity (ADMET). ADMET Prediction: Machine learning models can forecast a drug's potential toxicity as well as how it will be absorbed, distributed, metabolized, and eliminated from the body. Byidentifying and removing unsuitable drug candidates early on, these predictions help save time andmoney. Clinical Trial Optimization: By determining which patient subgroups are most likely to benefitfrom a drug, machine learning can enhance the planning and conduct of clinical trials. Trials becomemore effective and fruitful as a result.

Additionally, real-time clinical trial data analysis using machine learning models enables prompt detection of potential issues and adaptable fixes. Predictive modeling reduces costs and improves the chances of success by ensuring that only the most promising drug candidates go to clinical trials. In conclusion, machine learning is transforming drug discovery by lowering costs and increasing productivity. From target identification and validation to drug design, virtual screening, and clinical trials, artificial intelligence (ML) offers creative solutions to enduring issues in the pharmaceutical industry. As processing power and data availability rise, machine learning will become increasingly integrated into drug discovery, speeding up the creation of innovative, effective medicines. Collaborationbetween biologists, doctors, and computational scientists will be crucial to maximizing their potential.

Collaboration between biologists, doctors, and computer scientists is necessary to fully utilize machine learning in drug discovery.

Application

- *InSilico Models*: By simulating a drug's behavior in biological systems, machine learning models havethe potential to do away with the requirement for animal testing.

Best Practices for the Development and Validation of Ml Models

Data Organization and Selection: Every machine learning experiment begins with data curation on the training data required to build a model because a model is only as good as the data it is constructed from. Modelers typically employ data created (and, more recently, collated) by other scientists. As a result, the data sources often have a significant influence on the data quality. Although they are occasionally overlooked, data curation and preparation are crucial to building reliable models. Common issues with data curation and preparation are listed below.

Outliers

Observations that don't seem to match the rest of the data could be the consequence of data entry or experimentation errors. Finding anomalies and either removing or correcting them is one part of data curation. The modeler can also perform a sensitivity analysis by running the model again without the outliers and comparing the model results. Due to the size of many ML datasets, it may not be able to manually detect and manage outliers. It may be possible to visually analyze the distributions and summarize the covariate ranges in order to identify outliers or the ranges that contain them. Heuristic or statistical/mathematical approaches can be used to detect and handle data outliers in a semi-automated manner.

Missing Data

Another challenge in data preparation is handling missing data. Missing data is common; it often results from lost data entry or collection, as well as issues with data processing. Most machine learning algorithms have trouble with incomplete input, but some, like random forest, can function well with it. Incomplete data can be handled in a number of ways. The first steps should be to visualize the distribution of missing values and count the number of them. VIM and Missingno are two R and Python packages that can help with this operation. Generally speaking, variables with 70% or more missing data should be eliminated, depending on the variable's significance, as they often provide no insight into predictability.

Eliminating any data points with missing covariates is the simplest method for handling the remaining covariates. Using the median value for continuous variables or the mode for categorical variables in placeof missing data is another simple solution. For time-dependent variables, comparable statistical observations could be considered at various time points. These strategies make it possible for many approaches to fit machine learning models, but they also carry the risk of introducing bias. A more sophisticated technique for dealing with missing data for time-dependent variables is model-based imputation, like the regression model of children's height and weight. An additional model-based imputation method is the K-nearest neighbor (KNN) methodology. It is expected when utilizing KNN that the values of the non-missing points closest to a missing point value can be used to approximate that value.

Several effective techniques for managing missing data in well-established pharmacometric analyses have been documented in the literature. Handling missing data is usually quite challenging, and solutions must be used within the constraints of the particular problem. The majority of the time, effective missing data management may require a mix of several strategies.

Best Approaches for Curating Chemical Datasets for Drug Discovery

Chemical databases must be cleared of duplicate entries, and the suitability and correctness of the molecular representations must be confirmed. The degree of curation required depends on the molecular predictors that will be used in the models. For example, if only conformation-independent descriptors are to be used, geometry optimization or worries about the stereochemistry of the training samples may be avoided. The opposite is true if methods that consider molecular geometry are applied. Although it would be ideal to combine these automated procedures with manual curation, the modeler may employ cheminformatic tools to automate some of the earliest stages of curation. According to research, every work on the subject usually has two structural errors.

In 2008, Young et al. searched six commercial and public sources to obtain structural information on 8530 distinct chemicals. Observing that the error rates in the databases varied from 0.1 to 3.4%, they concluded that even small inaccuracies in structural representations can lead to significant errors in predictions. Maeda et al. recently searched the Natural Product Updates 2010 and 2011 database of the Royal Society of Chemistry. One by one, the publications from the designated literatures that reported newly identified or structurally updated natural substances were analyzed. Three percent had the correct name but the wrong structure, 0.4% had chiral atoms that were poorly portrayed in their drawings, and 63% of the about 2.5K compounds that required to be verified in the original source.

Techniques and Algorithms in Drug Discovery Machine Learning

Machine learning (ML) offers a wide range of techniques and algorithms that can be applied to various stages of the drug discovery process. An overview of several significant techniques and algorithms in thisarea is provided below:

Supervised Learning

Supervised learning involves training a model on a labeled dataset, where each training example has anoutput label. Common algorithms include:

- *Drug response levels* are one example of a continuous (linear) or binary (logistic) result that can bepredicted using regression analysis.
- *SVMs, or support vector machines*: Excellent for high-dimensional space classification problems suchas active compound prediction.
- *Random Forests*: An ensemble learning method for classification and regression that can be applied topredict drug-target interactions and ADMET features.

Basic feed forward networks for a variety of drug property prediction applications are included in neural networks.

Unsupervised Learning

When working with unlabeled data, unsupervised learning searches for hidden patterns or underlyingstructures. Common algorithms include:

- *Clustering Techniques*: Algorithms such as K-means, hierarchical clustering, and DBSCAN are used to group related substances or biological data.
- *Principal component analysis, or PCA:* A valuable technique for visualizing high-dimensional biological data that simplifies the most important parts of complex data is dimensionality reduction. Autoencoders are neural networks that are used for unsupervised feature learning and dimensionalityreduction.

Semi-Supervised Learning

Semi-supervised learning uses both labeled and unlabeled data to improve learning accuracy. In drugdiscovery, where there might not be a lot of labeled data available, this is extremely useful.

- *Extensions for Support Vector Machines (TSVM):* Because of this, SVM may use both labeled andunlabeled data.
- *Techniques Based on Graphs*: For molecular graph data, it is useful to distribute label information usingthe graph-based data structure.

Reinforcement Learning

Reinforcement learning encourages agents to make decisions in a specific order by rewarding desired outcomes. It aids in improving medication design:

- *Q-Learning* is an algorithm that calculates the value of an action in a particular state. Deep Q-Networks (DQN), which integrate Q-learning and deep neural networks, are useful for challenging drug design decision-making tasks.

Deep Learning

A subfield of computer learning called "deep learning" uses multi-layered neural networks to simulate intricate patterns:

- *CNNs, or convolutional neural networks*: Excellent in processing picture data, useful for deciphering biological images, and able to predict molecular properties from graphical representations.
- *RNNs, or recurrent neural networks:* RNNs are designed for sequence data and can be used to analyze protein sequences and drug interaction pathways.
- *Graph Neural Networks (GNNs):* Using graph data, like molecule structures, to predict attributes and interactions.

Generative Models

Generative models are used to generate new data points that share traits with the training data. Whendeveloping innovative medicinal molecules, this is beneficial.

- *GANs, or Generative Adversarial Networks*: consists of two networks (a discriminator and a generator) competing with one another to create synthetic data that seems realistic.

By determining the distribution of the training data, variational autoencoders, or VAEs, are used toproduce fresh data.

Transfer Learning

Transfer learning is the process of applying a previously learned model to a new, related activity, which cansave time and money:

- *Deep Learning Models* that are already trained: Large datasets (like ImageNet) can yield models that have been optimized for specific drug discovery tasks, such as chemical property prediction.

Natural Language Processing (NLP)

Large volumes of unstructured data and biomedical literature are processed and analyzed using naturallanguage processing (NLP) techniques:

- *Recognition of Named Entities (NER):* identifies and classifies textual entities, including medication andgene names.

Information retrieval and text mining extract relevant information from databases and scholarlypublications.

Applications in Drug Discovery

Target Identification and Validation: Using machine learning to analyze biological data in order to identify potential medication targets. Physical screening may not be necessary if virtual screening and hit findingare used to predict which chemicals will interact with a target. The process of enhancing a compound's chemical structure to boost efficacy and reduce toxicity is known as lead optimization. Predicting ADMET qualities involves examining a compound's absorption, distribution, metabolism, excretion, and toxicity. Clinical trial optimization is the process of using machine learning to design successful trials, identify suitable patient groups, and monitor safety and efficacy.

Additional Further Advance

Because machine learning algorithms use models that are far more complex than traditional statistical models, interpretability may be constrained. In certain situations, how the model is seen may depend on the relative importance of the covariates. This could be determined, for example, by calculating the model's accuracy reduction following a refit without a certain covariate. Another way to analyze the impact of a covariate is to plot the predicted outcome against the covariate while setting the other covariates to their respective dataset medians. In pharmacometric analysis, models are commonly used to simulate new data.

The simulation procedure is made more difficult by the fact that many machine learning models do not incorporate the outcome variability as a model parameter. If more simulated outcome data is requiredand the covariates are available, one approach is to begin with the model predictions and then add random noise with a variance determined by the model accuracy for the test or validation dataset. Nevertheless, when repeated measurements are made within subjects, this approach might not worksince model accuracy by itself is unable to differentiate between intra- and inter-subject variability. A course on the conceptual and technical aspects of machine learning was part of our first endeavor. Understanding this will help the reader comprehend our study's second section, which gives a summary of the applications of machine learning in the post-approval.

Challenges and Future Directions

Despite the promise, there are a number of obstacles to overcome: Integrating different data formats and providing comprehensive, high-quality datasets are examples of data quality and integration. Interpretability: Enhancing ML models' interpretability to promote regulatory acceptance and scientific comprehension. Scalability: Developing methods that function effectively with complex models and large datasets. Future paths consist of: Develop techniques to make explainable AI (XAI), or machine learning models, easier to understand. High-Throughput Technology Integration: ML in conjunction with high-throughput screening methods and genomics. Multidisciplinary Methods: Integrating machine learning with other scientific disciplines to take a comprehensive approach to drug discovery. Machine learning is transforming the drug discovery process by improving prediction accuracy, reducing costs, and speeding up the creation of new medicines.

Conclusion

This investigation has shed light on the revolutionary potential of machine learning (ML) in drug discovery,demonstrating its capacity to improve every stage of the drug development process, from lead optimization to target identification. As demonstrated, ML applications have improved the precision and effectiveness ofdiscovering and creating therapeutic molecules in addition to speeding up previously time-consuming procedures. This combination of science and technology signals a positive change toward more focused and effective healthcare solutions and provides hope for the quick creation of life-saving medications.

But there are obstacles in the way of incorporating machine learning into drug research, such as the requirement for large, high-quality datasets and the necessity of addressing AI ethics. The possibility of more developments as we approach this fascinating field encourages cooperation in honing AI applications so they enhance the priceless insights offered by human experience. In addition to revolutionizing drug development, the future of customized medicine and healthcare innovation is expected to be ushered in by the intersection of machine learning and pharmaceutical sciences.

Reference

Aspuru-Guzik, A., & Adams, R. P. (Eds.). (2020). *Machine learning for chemical discovery*.

Bajorath, D. (Ed.). (2018). *Machine learning methods in chemoinformatics*. Royal Society of Chemistry.

Gawehn, E., & Hiss, J. A. (2018). Deep learning in drug discovery. *Molecular Informatics, 37*(1-2), 1700153.

LeCun, Y., Bengio, Y., & Hinton, G. (2015). Deep learning. *Nature, 521*(7553), 436-444. https://doi.org/10.1038/nature14539.

Ragoza, M.,Hochuli, J., Idrobo, E., Sunseri, J., & Koes, D. R. (2017). Protein–ligand scoring with convolutional neural networks. *Journal of Chemical Information and Modeling, 57*(4), 942-957.

Schneider, P., Walters, W. P., & Plowright, A. T. (2020). SURE: A probabilistic model-based method forpredicting the success of drug discovery. *Nature Reviews Drug Discovery, 19*(6), 357- 358.

Zhou, Z., Kearnes, S., Li, L., Zare, R. N., & Riley, P. (2018). Optimization of molecules via deep reinforcement learning. *Scientific Reports, 8*(1), 1-10.

Chapter 3

The Role of Computational Methods in Drug Discovery

**B. Aarthi Rashmi*
N. Deesha Jayasri
and V. Perinbanayagi**
Department of Bioinformatics, Sri Krishna Arts and Science College, Coimbatore, Tamil Nadu, India

Abstract

Drug discovery greatly benefits from computational methods, which also drastically reduce costs. These techniques leverage large genomes and proteomics databases to aid in target discovery and validation. Molecular modeling, docking, and virtual screening predict drug-target interactions, which makes it easier to create compounds with the best possible attributes. Computational techniques are employed to forecast pharmacokinetics, pharmacodynamics, and toxicity, this helps in improving the efficacy and safety profiles of lead compounds. Structure-based drug design makes use of information from targets and three-dimensional structures to create exact medications. Biological activity is predicted from chemical structures using Quantitative Structure-Activity Relationship (QSAR) modeling, which directs lead alterations to maximize activity and reduce side effects. Predictive ADME/Tox approaches decrease late-stage failures by predicting the Absorption, Distribution, Metabolism, Excretion, and Toxicity features of drug candidates. Network pharmacology and systems biology resemble biological systems to comprehend medication, spot off-target effects and

* Corresponding Author's Email: aarthirashmib@skasc.ac.in

In: Harnessing AI for Drug Discovery
Editor: Kolla Bhanu Prakash
ISBN: 979-8-89530-540-9

find novel therapeutic applications. Data mining and machine learning are used in cheminformatics and bioinformatics to examine chemical and biological data and find patterns that help drug discovery. Molecular dynamics is one simulation technique that sheds light on the dynamic behavior of biological molecules, helping to understand drug resistance and mechanisms of action. Computational techniques also help high-throughput screening (HTS) by assisting with data analysis, compound prioritization, and identifying hit compounds that validate further development. The integration of various computational methodologies enables an efficient, inventive, and targeted drug discovery process, resulting in the development of safer and more effective treatments.

Keywords: drug discovery, computational methods, target identification, lead optimization, structure-based drug designing, ligand-based drug designing

Introduction

The drug discovery process is a challenging and time-consuming process focused on discovering new, prospective medicines. Traditionally, this approach has relied on substantial empirical research, along with considerable laboratory work and experimentation. However, the development of computational techniques has wholly transformed the drug discovery landscape, increasing its efficacy, cost-effectiveness, and efficiency. Computational methods are a significant step in the ever-changing medical research field. These latest techniques represent an important breakthrough in the search for novel therapies, optimizing the lead optimization process, improving targeting mechanism accuracy, and making it easier to identify promising compounds. Scientists require computational technologies since they accelerate the transition from an idea to an application. This method of drug design enables researchers to identify the compound's bio-targets. In reality, these methods will proceed towards producing safer drugs and effective medications.

Evolution of Computational Method

Several historical turning events took place for the beginning of computer-aided drug discovery or CADD, and it helped the field to grow over time. In

1981, an article known as Fortune magazine first highlighted the potential of CADD to transform the drug design by reducing the number of substances needed for screening while maintaining effective drug lead compound discovery. The more profound interest in the pharmaceutical industry and substantial financial commitment to computational methods has bloomed. The onset of CADD is closely related to advances in several scientific and technological fields. The advancement of knowledge in X-ray crystallography has significantly impacted molecular modeling and computational chemistry. Another prominent contribution is technological advancement in computers, which provides the computational power needed to process vast datasets and perform complex computations crucial for drug discovery.

In the 1890s, Emil Fisher proposed one of the earliest and most influential theories in drug design by linking drug-receptor interaction to an essential fitting into a lock. Daniel Koshland later advanced this concept by proposing the induced fit theory, which states that both the drug and the receptor undergo conformational changes during their interaction to achieve optimal binding. Numerous studies have validated this theoretical background, making it integral to modern computational drug design strategies. Another critical milestone in the use of computational methods is the development of in-silico modeling technologies, which allow for virtual screening, compound design, and analysis of drug-target interactions, all of which are crucial for the efficient selection of promising compounds.

Furthermore, applying machine learning methods to analyze structure-activity relationships represents a notable advancement in computational drug discovery. Techniques such as neural networks, random forests, and decision trees have significantly improved the predictive capabilities of computational models, making them far more effective than traditional statistical analyses. The historical development of computational methods in drug discovery has been marked by a series of technological and theoretical advances that have gradually enhanced the efficiency and accuracy of drug design.

These methods continue to evolve due to constant advancements in computational capabilities and an improved understanding of molecular interactions at the atomic level. Some tools and techniques that fall under computational methods include bioinformatics, molecular modeling (such as docking), cheminformatics, and machine learning. These can be used at any stage of drug discovery, from target identification to validation in clinical trials.

Target Identification and Validation

The initial step in drug discovery is identifying and validating biological targets, such as proteins or genes, that are associated with a specific disease. Computational methods are involved in this step of genomic and proteomic data analyses. The analyses of vast amounts of data generated by high-throughput techniques like NGS use bioinformatics tools that identify potential targets. These bioinformatics tools pinpoint mutations, expression patterns, and other anomalies in genes leading to disease. Systems biology approaches utilize computational models to elucidate the complex interactions among components within biological systems. Cellular pathways and networks allow for the creation of detailed maps that predict how interventions at specific points will impact the entire system. This approach helps identify key nodes for drug targeting.

Lead Compound Identification

Lead compounds are chemical substances used to treat diseases and can also facilitate new drug discovery. Several methods are involved in identifying lead compounds, including random and non-random screening, high-throughput screening, structure-based drug design, molecular docking (MD) simulation, and virtual screening.

Virtual Screening

Virtual screening is a computational technique used to discover potential lead compounds from extensive chemical libraries. It can be divided into two main approaches: structure-based virtual screening and ligand-based virtual screening.

1. *Structure-Based Virtual Screening (SBVS):* SBVS is a computational approach used in the early stages of drug discovery to identify a compound that binds to a drug target. This method involves docking small molecules into the active site of a target protein whose 3D structure is determined. An advanced algorithm simulates the

interaction between the ligand and the protein to predict how well the ligand fits into the active site and to determine the binding affinity.

2. *Ligand-Based Virtual Screening (LBVS):* Researchers use LBVS when the structure of the target protein is not determined. It mainly depends on the knowledge of other molecules known to bind to the target. Computational tools are used to identify new compounds with similar characteristics by analyzing the chemical features and properties of the molecules.

Lead Optimization

After identifying potential lead compounds, researchers must optimize them to enhance their efficacy, selectivity, and pharmacokinetic properties while minimizing toxicity. Computational methods facilitate this process through various techniques.

1. *Quantitative Structure-Activity Relationship (QSAR):* QSAR models utilize statistical and machine learning techniques to correlate the chemical structure of compounds with their biological activity. This approach enables the modeling of the activity of new compounds and the optimization of their chemical structures to enhance efficacy by *analyzing these relationships.*
2. *Molecular Dynamics Simulations*: MD simulations can generate information about the dynamic behavior of drug molecules and their interactions with biological targets in the form of time-dependent functions. Such evaluations provide insights into the reliability of a drug-target complex, potential binding modes, and conformational changes.

Admet Prediction

Absorption, Distribution, Metabolism, Excretion, and Toxicity (ADMET) properties are critical determinants of a drug's success. Computational methods can predict these properties early in the drug development process, thereby reducing the likelihood of late-stage failures.

1. *In Silico ADMET Models*: In silico ADMET models utilize chemical structures and employ algorithms and databases to predict ADMET properties. These models learning, quantitative structure-property relationships, and molecular docking. Researchers can swiftly conduct virtual screening of vast compound libraries to select those with a suitable ADMET profile, ensuring good oral bioavailability, low toxicity, and proper metabolic stability. In silico ADMET models also assist in prioritizing compounds for further experimental testing and save much time during drug development.
2. *Pharmacokinetic Modelling*: Pharmacokinetic models describe the drug absorption, distribution, metabolism, and excretion processes in the body. Setting up mathematical equations will further the ability of such models to explain the time course of drug concentrations in tissues and fluids. Once integrated with preclinical and clinical study data, the pharmacokinetic models predict the behavior of the drug in other patient populations with various physiological or pathological conditions. Pharmacokinetic modeling helps select appropriate dosing regimens and thus predicts possible drug interactions but avoids specific drug-drug interactions due to genetic variations in drug metabolism. This advice will be relevant for launching a new drug on the market, ensuring that it is both safe and effective.

Drug Repurposing

Drug repurposing refers to the process of assigning new therapeutic uses to drugs that are already available on the market. This process heavily relies on computational approaches, as they enable the rapid identification of the potential efficacy of existing drugs for new therapeutic applications.

1. *Computational Drug Repositioning*: It involves screening already available drug and disease relevancy databases for new indications for already existing drugs. It therefore utilizes computational approaches such as data mining, machine learning evaluations, and network analysis. Computational approaches relate to an architecture developed from the disease indications and molecular and phenotypic attributes of currently approved drugs to generate an evaluation of previously unknown associations that may aid in new therapeutic pathways for currently approved drugs. Rather than developing new

drugs from scratch, this model saves time and money while limiting the potential for adverse effects since the drugs repurposed.

Personalized Medicine

The goal of personalized medicine is to develop treatment adjustments based on each patient's predispositions, which are influenced by the expression of specific genes, proteins, or metabolic patterns. Computational methods play a crucial role in integrating and analyzing the extensive data needed to create personalized treatment plans.

1. *Pharmacogenomics:* Pharmacogenomics studies how genetic variations affect individual responses to drugs. Computational tools analyze genetic data to predict how patients will respond to specific treatments, enabling personalized drug therapy. Pharmacogenomics is used to choose the appropriate drugs and their dosages for individual patients by identifying genetic polymorphisms that influence drug metabolism, effectiveness, or toxicity. This approach aims to enhance therapeutic efficacy while minimizing side effects, ultimately leading to more effective and safer treatments.
2. *Computational Biomarkers*: Biomarkers are measurable indicators of biological processes or diseases. Computational methods can identify and validate biomarkers from omics data, facilitating the development of personalized treatment strategies. By analyzing large datasets from genomics, proteomics, and metabolomics studies, computational tools can discover biomarkers that predict disease risk, progression, and response to treatment. Subsequently, these biomarkers can be used to stratify patients, monitor diseases, assess progression, and evaluate therapy efficacy, thereby contributing to more accurate and personalized healthcare.

Artificial Intelligence and Machine Learning

Artificial intelligence has transformed various areas of drug discovery. These methods can analyze and process massive data volumes, detecting patterns and insights that humans cannot observe.

1. *Deep Learning in Drug Discovery*: Deep learning is a subset of machine learning that focuses on neural networks with multiple layers capable of understanding complicated data patterns. In drug discovery, researchers utilize deep learning models to perform tasks such as predicting molecular properties, identifying potential drug candidates, and designing new molecules. For instance, they can train these models on large datasets containing chemical structures, biological activities, or experimental results to recognize specific patterns that may indicate promising drugs. The approaches under deep learning can also help optimize the chemical structure of lead compounds while assessing their ADMET characteristics and possible off-target effects.
2. *Natural Language Processing (NLP):* NLP techniques allow valuable information to be extracted from scientific articles, patents and clinical trial reports. NLP assists researchers in staying updated with the latest discoveries by automating the literature review process, thereby aiding their research related to new drug targets and compounds. NLP algorithms can analyze significant amounts of data and extract information about gene-disease associations, mechanisms of action of drugs, and outcomes of clinical trials. NLP algorithms can analyze large bodies of text to extract relevant information.

Integration of Multi-Omics Data

The integration of data from diverse omics technologies (genomics, proteomics, metabolomics, etc.) enables a comprehensive understanding of biological systems and the causes of diseases. This approach incorporates computational methods to analyze and interpret multi-omics data in drug discovery.

1. *Network Pharmacology*: Network pharmacology refers to understanding the actions of drugs at the level of complex biological networks rather than at a single target. Computational tools help construct and analyze these networks, identifying multi-target strategies for complex diseases. By mapping the interactions between drugs, targets, and biological pathways, network pharmacology provides insights into the polypharmacology of drugs and their effects on entire biological systems. This approach further offers

opportunities to unravel synergistic effects, identify potential side effects, and provide direction for combination therapies against multifactorial diseases.

2. *Systems Pharmacology Models*: Pharmacology models integrate data from different biological levels to predict the effects of drugs on the entire organism. They employ computational simulations to analyze drug-biological system interactions, considering genetic variations, environmental influences, and disease states. Computational models in pharmacology illustrate how drugs interact with one another, their side effects, and how individual genetics affect drug responses. They can also be used to predict the therapeutic and adverse effects of new drugs, optimal dosing regimens, and to guide clinical trial design.

Challenges and Future Directions

The challenges and future directions of integrating the computational methods in drug discovery are Data Quality and Integration, Model Interpretability, Ethical and Regulatory Consideration

1. *Data Quality and Integration*: One of the major challenges in computational drug discovery is the quality and integration of data from numerous sources. To achieve accurate predictions and insights in drug discovery research, it is essential to standardize data formats, enhance curation processes, and develop interoperable databases, as these measures facilitate data exchange and integration.
2. *Model Interpretability*: AI and ML models are being used for their powerful predictive capabilities, but interpreting their predictions is primarily challenging due to their black-box nature. Currently, researchers are focusing on developing interpretable models that provide insights into the underlying biological mechanisms. Techniques such as Explainable AI (XAI) enable researchers to trust and validate their predictions while gaining a deeper understanding of the biological processes they are studying. This approach also aims to make machine learning models more understandable and accessible.
3. *Ethical and Regulatory Considerations*: In drug discovery, the increasing use of computational methods has raised ethical and regulatory considerations, particularly concerning data privacy and

the validation of computational predictions. To successfully implement these approaches, it is essential to ensure that they adhere to ethical standards and regulatory requirements. Addressing these challenges involves developing guidelines and frameworks for the ethical use of AI and ML in drug discovery, safeguarding patient data privacy, and rigorously validating computational models.

Conclusion

Integration of computational methods has transformed drug discovery by skyrocketing the processes from target identification to personalized medicine. By analyzing genomic and proteomic data, these advanced tools identify disease-associated targets, streamline lead compound discovery through virtual screening techniques, and optimize drug candidates using models like QSAR and molecular dynamics simulations. Additionally, predicting ADMET properties enhances candidate selection, while computational approaches facilitate drug repurposing by identifying new therapeutic uses for existing drugs. The precision and efficacy in developing safer and more effective treatments across diverse medical conditions are enhanced because of the integration with experimental methods, which marks the computational techniques as pivotal in shaping the future of pharmaceutical research and development.

References

Kaur P, Khatik G. An Overview of Computer-aided Drug Design Tools and Recent Applications in Designing of Anti-diabetic Agents.

Moses, H., Dorsey, E. R., Matheson, D. H., & Thier, S. O. (2005). Financial anatomy of biomedical research. *Jama*, *294*(11), 1333-1342.

Myers, S., & Baker, A. (2001). Drug discovery—an operating model for a new era. *Nature Biotechnology*, *19*(8), 727-730.

Razia IT, Kanwal A, Riaz HF, Malik A, Ahsan M, Khan MS, Raza A, Sabir S, Sajid Z, Khan MF, Tahir RA, Sehgal SA. Recent Trends in Computer-aided Drug Design for Anti-cancer Drug Discovery.

Richards WG. Computer-aided drug design. Pure Appl Chem 1994; 66:1589-96.

Sabe, V. T., Thandokuhle Ntombela, Lindiwe A. Jhamba, Glenn E. M. Maguire, Thavendran Govender, Tricia Naicker, Hendrik G. Kruger. Current trends in

computer-aided drug design and a highlight of drugs discovered via computational techniques: a review. *Eur. J. Med. Chem.* **224**, 113705.
Skjevik, Å. A., Teigen, K., & Martinez, A. (2009). Overview of Computational Methods Employed in Early-Stage Drug Discovery. Future Medicinal Chemistry, 1(1), 49–63.
Terstappen GC, Reggiani A. In silico research in drug discovery. *Trends Pharmacol Sci* 2000; 22:23-6.

Chapter 4

Virtual Screening and Compound Prioritization

Vidya Niranjan*
Spoorthi R. Kulkarni
Rao Deeksha Umesh
and Shradha Anand Mulimani
Department of Biotechnology, RV College of Engineering,
affiliated to Visvesvaraya Technological University, Belagavi, India

Abstract

The identification of attractive therapeutic leads from vast chemical libraries is a critical and growing aspect of drug discovery. Virtual screening (VS) and compound prioritization have become indispensable in this endeavor, utilizing in silico techniques to assess the biological activity of compounds and expedite development processes. VS encompasses methods to analyze large compound libraries against specific biological targets, enabling the computational study of binding interactions between small molecules and proteins. This facilitates the filtering of less promising candidates, concentrating experimental efforts on more productive leads, thus accelerating hit identification and reducing costs. VS is categorized into ligand-based and structure-based approaches. Ligand-based virtual screening uses the chemical structures and properties of known active compounds to identify new, potentially similar molecules. Structure-based virtual screening leverages the 3D structures of target proteins to find ligands that fit their active sites, employing docking simulations and scoring functions. The VS workflow

* Corresponding Author's Email: vidya.n@rvce.edu.in

In: Harnessing AI for Drug Discovery
Editor: Kolla Bhanu Prakash
ISBN: 979-8-89530-540-9

involves library preparation, target identification, target modeling, docking simulations, scoring, and post-processing. Compound prioritization ranks shortlisted compounds based on efficacy, selectivity, safety, pharmacokinetics, drug-likeness, synthesis feasibility, and intellectual property considerations. This step ensures that selected candidates possess high biological activity and favorable drug-like properties. Efficient compound prioritization integrates VS results with cheminformatics and machine learning models to identify molecules for experimental validation. This chapter covers the principles and methods of virtual screening and compound prioritization, exploring ligand-based and structure-based approaches, molecular docking, and integration with machine learning. It also discusses multi-parameter optimization and the integration of computational and experimental data, essential for progressing quality compounds in drug discovery pipelines.

Keywords: virtual screening, high-throughput screening, ligand-based virtual screening, structure-based virtual screening, molecular docking, machine learning, compound prioritization

Introduction

The identification of attractive therapeutic leads from vast chemical libraries is critical in the field of drug discovery, which is ever-growing. Today, the use of virtual screening (VS) and compound prioritization represents quite indispensable fingerprints to unravel such mysteries that utilize in silico techniques to categorize the biological activity of compounds and expedite the development process [1]. Virtual screening is a collection of in silico methods used to analyze compounds from large compound libraries against specific biological targets. VS allows for the computational study of individual binding interactions between small molecules and a protein target so that less likely candidates can be filtered out and experimental efforts can be concentrated on more productive leads [2]. This method not only speeds up hit identification but also greatly lowers the costs and time spent in the first phase of drug discovery [3]. There are two main types of virtual screening: ligand-based and structure-based. Ligand-based virtual Screening uses the chemical structures and properties of known active compounds. Ligand-based screening identifies new, potentially similar molecules that may exhibit comparable biological activity [4]. Prediction methods, such as quantitative structure-activity relationship (QSAR) modeling and similarity searching, have been established to infer the efficacy of new ligands. On the other hand, structure-based virtual screening exploits to its advantage the three-dimensional (3D) structures of target proteins and searches for ligands that might fit in their active sites [5]. A method of screening a small molecule library is used to predict which drugs will most effectively bind the target protein. First, either a 3D protein model or an X-ray crystallographic structure of the protein complexed with a bound ligand is subjected to grid generation. The binding sites are then docked with small molecules, and computational scoring functions evaluate their ability to bind, along with other characteristics.

The process and workflow of virtual screening consists of several key steps. It starts with a library prep, in which we take a huge number of small molecules and prepare them for screening. Target identification is then followed by targeting modeling, which selects and prepares the target protein structure [6]. Following identification of the target binding site isolated is docking simulations, which are utilized to computationally mimic sets up between small molecules and the corresponding protein. Scoring and ranking - it quantifies the binding affinities of the docked ligands and orders them according to their respective scores [7]. In the end, post-processing goes to work on how likely the best hits are a drug and what else they could bind.

Compound prioritization is when one ranks shortlisted compounds based on multiple attributes, which include evaluating efficacy (potency and effectiveness), selectivity (targeting specific proteins or pathways), safety/toxicity profiles, pharmacokinetics/pharmacodynamics (ADME and drug concentration effects), drug-likeness, synthesis feasibility (complexity/cost), and intellectual property considerations [6]. These elements are key inputs when deciding to take compounds through drug development. This step is important in making sure that the candidates selected have not only high biological activity but also good drug-like properties. Efficient compound prioritization combines results obtained from virtual screening, cheminformatics approaches, and Machine Learning models in order to identify a set of molecules that should be taken forward for experimental validation [7]. The scientific tools and technologies available for drug discovery include high-throughput screening (HTS) to rapidly test large numbers of potential compounds, computational modeling that predicts interactions between ligands/targets at the molecular level through computer simulations with biological assays which evaluate activity in vitro or ex vivo together with chemoinformatics [8] useful in processing data. Together, these tools facilitate rapid pre-clinical compound selection by binding affinity, biological activity and multi-parameter ranking systems for progression of quality compounds in drug discovery pipelines. This chapter covers principles and methods of virtual screening, as well as compound prioritization. It explores the different types of virtual screening that will be examined, from ligand-based to structure-based approaches, focusing on their applicability and potential issues, such as the principles of molecular docking and the integration with machine learning. It will also discuss how to prioritize compounds and focus on multi-parameter optimization together with the consideration of computational/experimental data integration.

High-Throughput Screening (HTS) Data Analysis

High-Throughput screening (HTS) is a method commonly used in biological and drug discovery research that allows the assay of as many as hundreds of thousands of compounds for an activity. It is a crucial step in drug discovery where screening of compound libraries is used to find hits or leads that act on the target in an appropriate way [9]. Through this method, one can rapidly screen many excipients for affinity towards a target protein by completely automated means (either calling the ligand activators or inhibitors of binding affinity). This allows researchers to discover new drugs more efficiently, making it quicker and easier for them to find effective treatments [10]. High throughput screening (HTS) for compound identification is well-established and universally practiced in modern pharmaceuticals, such as the biotechnology sector. HTS is an important tool for evaluating a large number of molecules, and in vitro models may represent a suitable method. The assays are run in 96, 384 or even more recently developed ultra-high-density microtiter plates [11, 12]. The use of miniature consumables allows reactions to be carried out in nanoliter-size volumes, providing an economical and expedition timescale-efficient solution. This reduction in scale requires fewer reagents and allows more compounds to be tested simultaneously. This is to identify compounds that can bind or interact with a desired biological target [13]. HTS relies on state-of-the-art robotic systems to process large volumes of samples and assays. These systems automate reagent dispensing, mixing, and liquid handling to a great extent, which not only decreases human intervention but also results in more throughput. High throughput screening (HTS) needs robotic platforms with the capability to work on multiple microtiter plates at one time so the operation continues and more samples can be processed at once [14]. A critical feature is to ensure reagents and samples are handled with the same precision on every occasion, as this must be cast iron before any screening data can be considered reliable or reproducible. Moreover, the combination with advanced software allows intelligent and adaptable control of all components throughout a fast-running screening process. The old traditional way was to test relatively few compounds by hand, which takes a long time and costs money, while HTS permits researchers to screen thousands to millions of compounds in a fraction of the time. This rapid throughput speeds up initial drug discovery and increases the probability that true hit compounds with therapeutic potential are identified. HTS also enables innovation by enabling the interrogation of thousands to millions of compounds and biological targets [15]. Their rapid profiling of diverse compounds can reveal new chemical scaffolds and molecules with novel mechanisms. Such diversity is vital in studying complex diseases and, subsequently, when designing therapeutics that attack different biological pathways or cellular processes. In addition to speeding up drug discovery, HTS provides a gain in overall efficiency and more economic and innovative solutions. High-throughput screening of several thousands to millions of compounds against specific biological targets generates large amounts of data, the magnitude and complexity of which necessitate advanced analytical methods. The role of data analysis is to quickly select already performed experiments relevant and interesting "hit" compounds, which demonstrates the desired biological activity in a way not covered by noise or false positives [16]. It also aids in structure-activity relationship (SAR) analysis to fine-tune and validate lead compounds for proceeding ahead with the development stage [17]. In addition to target identification, data-driven approaches enable a broad analysis of omic datasets that can reveal patterns, trends, and correlations in candidate compounds with potential mechanisms of action or undesired effects. It boosts the efficiency and hit rate for the discovery of leads through HTS, thereby becoming vital in linking experimental results and intelligible scientific progressions within pharmaceutical research.

In HTS, data analysis uses statistical tests to differentiate between real active compounds ("hits") and background noise (false positives) [18]. The use of thresholds and algorithms enables researchers to remove compounds that do not contain predefined characteristics for biological activity, saving time by focusing resources on potential candidates [19]. It is also important for interpreting structure-activity relationships (SAR). It is the linking of chemical structures for chemicals, proteins, and nucleic acids with biological activity or structural features. These features enable data-driven discovery of unexpected correlations or novel biological mechanisms. It helps in prioritizing compounds for subsequent verification and optimization, thus decreasing the prospects of false positives while enhancing the probability of identifying potential drug candidates. Data analytics is used to provide quantitative evidence and insights informing decision-making across the drug development pipeline. Acutus speeds drug discovery by helping researchers navigate through large datasets to identify exciting leads in the development of new therapeutic agents [20].

Data Collection

The most fundamental step in data analysis is the collection of raw data. The process of collecting initial experimental data from assays designed to test the biological activity of compounds against specific targets is called Raw Data Acquisition [21]. The efficiency and accuracy of data collection are paramount to the success of HTS, as they directly influence the quality and reliability of the subsequent data analysis and hit identification processes [22]. Raw data acquisition in biological and biochemical experiments involves various methods to measure signals. Methods such as fluorescence, luminescence, absorbance, etc., are commonly used [23]. Fluorescence is a rapid and sensitive method where a substance absorbs light at one wavelength (excitation) and emits light at a longer wavelength (emission). Used with fluorometers or spectrofluorometers, it labels DNA/RNA in sequencing for quantification and quality checks. It monitors processes like Illumina cluster generation and nucleotide flow, validating and quantifying sequences post-sequencing. Offering high sensitivity, it detects low concentrations accurately, binding specifically to targets in complex samples, versatile across molecules and systems, and vital in research and diagnostics [24]. Luminescence, the emission of light without significant heat, is detected using specialized instruments like luminometers or spectrofluorometers, measuring light intensity to gauge luminescent molecule activity or concentration. High-Throughput Sequencing (HTS) data analysis validates genomic findings via reporter gene assays (e.g., firefly luciferase) for gene expression and ATP assays for cellular responses, aiding drug discovery by screening compound libraries [25]. Luminescence assays excel with high sensitivity, specificity, and a wide dynamic range for diverse experimental needs, supporting real-time monitoring and non-invasive imaging in biological research. UV-Vis spectroscopy is essential in biological and biochemical research to quantify light absorbance by substances in solution. It measures the absorbance of UV and visible light by molecules, revealing their composition and concentration based on a electronic structure. UV-Vis spectrophotometers, equipped with UV (200-400 nm) and visible (400-700 nm) light sources, detect transmitted light intensity [26]. This method quantifies biomolecules like DNA, RNA, and proteins (DNA at 260 nm, proteins near 280 nm), monitors enzyme kinetics, and ensures HTS sample quality. It also characterizes biomolecular structures and interactions, aiding in understanding biological processes and drug development. UV-Vis spectroscopy offers high sensitivity (nanomolar to micromolar range), rapid measurements, and accurate quantification, which is crucial for diverse applications in molecular biology, biochemistry, and pharmaceutical sciences [27]. In addition to the diversity of methods for generating and analyzing HTS data, several specialized techniques play crucial roles in extracting detailed biological insights. DNA sequencing encompasses shotgun sequencing for comprehensive genome coverage, targeted sequencing for specific genomic regions, and whole genome sequencing for complete organismal genomes [28]. RNA sequencing (RNA-Seq) methods like mRNA-Seq and small RNA sequencing quantify gene expression and microRNA activity, respectively. Epigenomics investigates DNA methylation patterns and histone modifications that regulate gene expression, while ChIP-Seq identifies protein-DNA interactions to map genomic binding sites [29, 30]. Metagenomics analyzes microbial communities through 16S rRNA sequencing and shotgun metagenomics, offering insights into environmental and clinical samples. Long-read sequencing enhances genome assembly and transcriptome complexity analysis, while single-cell sequencing delves into cellular heterogeneity [31]. These methods collectively enrich HTS data analysis, enabling deeper biological understanding across diverse research fields. The data which has been collected needs to be handled carefully and precisely. This can be done with the help of Automated Systems. Automated systems are critical in High Throughput Screening (HTS) for efficient data collection, ensuring consistency and accuracy in drug discovery workflows. These systems precisely dispense compounds, reagents, and assay components into microtiter plates, minimizing variability and maintaining uniform assay conditions across multiple samples [32]. By reducing human error in pipetting and handling, automation enhances the reliability and reproducibility of experimental results. In addition to precision, automated systems significantly increase throughput by executing assays in parallel across multiple microplates, facilitating the rapid screening of thousands to millions of compounds. This capability accelerates the evaluation of large compound libraries, optimizing resource utilization and reducing operational costs associated with manual labor. Integrated with microplate readers and detection systems, automated handlers capture real-time assay signals such as fluorescence, luminescence, or absorbance. This integration streamlines data acquisition and ensures efficient monitoring of assay progress, contributing to a streamlined workflow from sample preparation to data analysis. Advanced informatics platforms support data organization, annotation, and preliminary analysis, facilitating hit identification and decision-making in drug discovery [33]. They improve experimental precision, reduce variability, and support rapid screening of compound libraries, ultimately advancing the identification and development of potential therapeutic candidates in drug discovery.

Data Preprocessing

Post-data acquisition comes with Data preprocessing. High-throughput sequencing (HTS) technologies generate millions of short DNA sequences in a single run to produce the whole genome sequence or for target loci [34]. Pre-processing is used to improve sequencing data by correcting biases and removing artifacts; this allows accurate comparison between samples as well as downstream analyses such as variant calling, genotyping, etc. They are: 1) normalization, 2) data QC (quality control), and 3) data cleaning. Normalization methods normalize HTS data with respect to sequencing depth and other technical artifacts that can change between samples [35]. Library Size Normalization adjusts the read counts to remove technical variation as much as possible so that what's left is only biologically pertinent. The values RPKM/FPKM (Reads/Fragments per Kilobase of transcript per Million mapped reads) and TPM (Transcripts Per Million) scale read counts by the length of a given transcript and how much it sequenced, giving reasonable estimates comparing gene expression in different samples [36]. Data reliability and excellent reproducibility are provided by the terms of quality control across the board with stringent metrics. The Z-factor compares the variability with positive and negative controls to the plate standard deviation. Signal-to-Noise Ratio (SNR) expresses the strength of a signal relative to background noise, which is very important for identifying real biological signals in contrast with technical noises. Coefficient of Variation (CV) A measure of data dispersion around the mean, and thus to assess precision/reproducibility for an assay [37]. Data cleaning means cleaning data by removing missing values, outliers, and noise so as to improve the integrity of data. Imputation is a way to replace missing data fields by mean/median, including column similarity, or using the K-Nearest Neighbors (KNN) method [38]. Statistical test-based outlier detection is used to find data points that are significantly different from the rest of the dataset, and it helps in robust statistical analysis. Background noise can be reduced by averaging or Gaussian smoothing and also using denoising tools such as low-pass filtering, differential processing (MCCMS), CHESS-VIPR imaging technique level-set approach in order to increase data quality like false positive/negative reduction [39]. Through these preprocessing steps, researchers are able to make the HTS data more rugged and worthwhile when it comes to producing strong biological conclusions. These practices are important for finding gene expression patterns, detecting variants, and conducting genomic analysis, which are necessary for progressing biological and biomedical research.

Hit Identification

The next most significant step is Hit Identification. Hit identification is the process of identifying lead compounds or targets from a pool of high-throughput screening data [40]. This includes in-depth statistical analysis and validation to separate the hits from the chaff of background noise and artifacts that appear coated over all high-throughput biological measurements. Selection of the hit threshold is a critical step in HTS. The thresholds to be applied are chosen depending on parameters like the assay sensitivity, desired hit rate, and available resources for follow-up studies. A common approach for hit selection is to use a percent inhibition threshold so that compounds with activities above a certain percentage in binary assays (e.g., a 50% decrease of the target activity) are chosen as hits [41]. The approach is simple and easy to understand; however, it does not cater to the dispersion in our assay. Alternatively, a Z-score threshold may be used such that compounds with a Z-score over some critical value (e.g., >3) are tagged as hits. A Z-score is a standardized measure that reflects the number of standard deviations by which an activity of a compound differs from the mean calculated on controls. It is more sensitive in identifying hits since it accounts for the assay variability [42]. Some researchers use a percent of control threshold, where compounds with activity above a certain percentage of the positive control are selected as hits. This method is useful when positive control is well-defined and represents the desired level of activity. Results from HTS can also be used to detect hits with a variety of statistical methods. For example, the activity of each compound relative to the negative control is expected to be compared using t-tests. If the activity of the two compounds differs more significantly (p-value < 0.05), we call them potential hits. One-way Analysis of Variance (ANOVA) can be used to analyze the variance across plates or experiments. This method may help to highlight the compounds that have consistent activity, and which are markedly distinct from controls. The fold-change analysis is also permitted, whereby researchers expect the compounds that achieve activity above a certain fold-change compared to the negative control hits likely to possess a significant effect size. The method is applicable when the assay has a wide dynamic range, and we anticipate that hits will have significant effect sizes [43].

Data Visualization

Once potential hits are identified from statistical analysis, researchers use various data visualization techniques such as heat maps, scatter plots, and box plots to visually investigate assay responses, detect dose-dependent effects within the datasets, and provide a visual summary of patterns or anomalies across compound libraries. Such visualizations help validate the preliminary results, scrutinize hits for robustness, and provide evidence-based guidance toward prioritization of drug discovery efforts [44]. Heatmaps are extremely powerful for visualizing large datasets, particularly allowing us to view patterns present in the assay results across many conditions or compounds. They show data in a grid of colored cells, where colors indicate changes between assay responses. Heat maps allow researchers to detect patterns, e.g., concentration gradients where responses increase from the bottom left corner of the plot moving towards the top right or grouping compounds that display similar activity profiles. It is especially powerful in high throughput screening (HTS) to identify hits or dose-dependent effects across a large number of samples [45]. In HTS, scatter plots are used to compare different variables, such as assay response versus compound concentration or assay responses, between experimental conditions. They represent data points as dots in 2D Cartesian space, with each axis representing a different variable. Scatter plots in HTS are valuable for highlighting any potential correlations, outliers, and trends that might be hidden using only aggregated data. Box plots are used in conjunction with scatter plots to portray a summary of distributions, offering statistical properties such as the median and quartiles and showing when outliers occur [46]. Box plots for HTS are used to show the variation of readout data summing multiple conditions or different. These are used to compare the spread and central tendency of data, which allows researchers to evaluate assay quality or lets them see if there is something off with the restyling results [47]. Interactive dashboards provide a better environment for data exploration in HTS because it is possible to dynamically visualize and interact with large datasets. Interactive Graphs provide you with the facility to create this type of graph such that if a user wants, he can even drill down into those visualizations. For example, in HTS, interactive dashboards allow the examination of assay results that can be drilled down to subgroups and filtered on the fly, exploring trends or outliers across various facets (e.g., compound libraries, assay conditions). These software tools to facilitate HTS data review include visualizations, allowing for collaborative decisions based on the analysis of a picture in combination with other numerical analyses. HTS depends on interactive dashboards to enable fast and efficient browsing of the input pool, thus promoting exploration and hypothesis, guiding forward or back steps in looking for leads, and queries about potential targets/endpoints that can be further explored [48].

Secondary Screening

After visualization of HTS data, such as heat maps or scatter plots, the hit activity was confirmed using confirmation assays and dose-response curves to validate their robust biological activity [49]. It is crucial during the validation and characterization of hits from the lead optimization stages after High Throughput Screening (HTS) in drug discovery. It is an ultra-important part of your preclinical studies and a critical phase for confirming the bioactivity of hits identified in primary screens as well as being able to quantify how potent/efficacious they are. Secondary confirmation assays are performed to verify the hits identified in primary HTS screening. Primary screens identify candidate compounds that act on the biological target or pathway of interest, whereas confirmation assays rescreen hits under increasingly stringent conditions. This phase is implemented in order to validate the reproducibility and specificity of top findings, filtering true hits from false positives or artifacts caused by experimental inconsistencies. Researchers validate hits by retesting them over a concentration series in confirmation assays. This not only confirms the results of this initial screening but also helps ensure that compounds are prioritized in drug development pipelines with confidence [50]. Dose-response curves are critical for the characterization of the potency and efficacy of lead hits found in HTS. These curves illustrate the relationship between the concentration of a compound and some effect (generally in an assay) caused by that chemical. Dose-response curves are generated for assays run in multiple concentrations [51]. All data points on the curve represent an average of how well the assay responded at one single compound concentration. Dose-response curves are used to determine IC50 (half maximal inhibitory dose) or EC50 (effective concentration of 50%) values. Namely, the IC50 and EC50 values denote, respectively, the tested concentration of substance necessary to inhibit 50% (IC) or induce a maximal effect corresponding to ~50-60% with respect to control [52]. As such, it serves as a well-characterized pharmacological profile that is essential for the evaluation of potency and optimization work. One of their purposes is to be used as a quality control measure for high throughput screening (HTS) dose-response curves. They are used by researchers to determine the reproducibility of assay results, conducting follow-up studies designed to confirm activity within hits identified in a primary screening while at the same time supporting compound robustness ability throughout downstream parts of drug discovery.

Data Integration

After hit confirmation in secondary screening and generation of dose-response curves to determine their potency/effectivity, the next stages include integration/annotation of data to place these hits into a chemical/biological context. During this step, chemoinformatics tools are instrumental in correlating chemical structures to their bioactivity profiles, providing insight into critical structural features modulating the observed effects. Virtual screening, along with the subsequent Structure-Activity Relationship (SAR) analysis, allows you to predict and optimize how these compounds will interact with their target, speeding up the search for novel drug candidates [53]. These tools are useful because bioinformatics is necessary for connecting annotated hits with biological pathways and targets. Among numerous pathway analysis tools, KEGG and Reactome facilitate the mapping of compounds to existing biological pathways (and, in turn, illuminate potential mechanisms underlying their effects) [54]. Several target prediction tools, such as DrugBank and BindingDB, can be employed to predict potential targets of compounds based on known interactions, physicochemical descriptors, or structural behavior in the further steps for experimental validation/characterization. HTS data can be seen as a lengthy series of wells with recordings, and integration into public or proprietary databases provides another dimension to the HTS results. Public databases (like PubChem and ChEMBL) provide large datasets regarding chemical properties and bioactivities to cross-reference the results of HTS so that they can be validated [55]. Pharmaceutical companies' proprietary databases hold sensitive data that could be used to generate more comprehensive insights and facilitate joint research. This data and annotation integration bring context to the work that researchers do on a discovery level across many fields, including chemistry and biochemistry. This holistic knowledge is very important to the efficient identification and optimization of leads in drug discovery.

Statistical and Computational Analysis

The integration of diverse datasets and extensive annotation offers a strong backbone for advanced statistical and computational analyses. Combining chemical, biological, and pharmacological information allows researchers to utilize more sophisticated multivariate analysis together with various machine learning methods to improve the identification of hidden signals and predict successful hits. These are important in analyzing the large-scale screening data resulting from HTS experiments, drawing significant patterns, predicting hits, and prioritizing compounds for follow-up [56]. Multivariate tools are essential for investigating multiple variables at the same time and searching relationships of high complexity in HTS data. This is a requirement to characterize the complexity of biological responses and chemical interactions. Principal Component Analysis (PCA) and hierarchical clustering are common techniques used. One such technique is PCA (Principal Component Analysis), which reduces a large set of variables to a small one, containing most information from the original [57]. PCA also serves as a visualization and interpretation tool in HTS, which can be used to summarize the most substantial variations of data, uncover patterns, or group compounds according to their biological activities. The latter kind of hierarchical clustering groups compounds in terms of their similarity in biological activity profiles and produces a dendrogram that is illustrative with regard to these relationships. It is especially useful for clustering compounds that evoke common effects and may point to some pathway mechanisms or unintended consequences [58]. Machine learning algorithms can then be used to predict hit compounds, and herein, we applied these machine-learning-based methods following on from the learnings established by multivariate analysis. When fed historical data, these algorithms are capable of making accurate predictions about brand-new compounds, significantly improving the efficiency of drug discovery [59]. Machine learning is crucial for the analysis of HTS data because it permits hit compounds to be predicted and ranked with different algorithms. Random forests are an ensemble learning method for classification, regression, and other tasks that operate by constructing a multitude of decision trees at training time and outputting the class or mean prediction (in the case of regression) of the individual trees [60]. Using a wide range of chemical and biological features, random forests predict compound activity robustly - thereby improving the hit identification process. Another wonderful tool in HTS is the Support Vector Machines (SVMs). The supervised learning models can be used for classification as well as regression tasks. Since SVMs group the molecules based on their properties as Active or Inactive, it makes it easy to identify drug candidates quickly and helps reduce the time for promising compounds by grouping them [61]. Among others, neural networks and especially deep learning models have large potential for the modeling of complex relationships in HTS data. This means that they can work with big data for many features, and we know that these methods are powerful in the quantitative structure-activity relationship, but this time the feature weight will be more important [62]. These machine learning methods altogether enhance the efficiency and accuracy of HTS data analysis, which in turn speeds up the discovery process for new therapeutic compounds.

Use of AI in High Throughput Screening

By providing advanced algorithms and computational powers, High-Throughput Screening (HTS) has influenced data analysis and hit identification in a very smart way with AI. AI-driven methods dramatically improve the speed and precision of HTS data processing, providing new perspectives on drug development. AI-powered data analysis uses high-level AI techniques such as natural language processing (NLP) and computer vision to process many different types of HTS datasets into findings. Using NLP algorithms, we can parse and analyze data in textual form (for example, scientific literature or assay reports) to extract information about compound activities and biological targets [63]. On the other hand, computer vision algorithms are powerful in processing pictures from high-content screening (HCS), identifying minor alterations in cellular morphology or protein localization provoked by compounds. In practice, this capability permits researchers to probe more detailed cellular mechanisms of action [64].

AI-powered predictive modeling helps to predict the activities of compounds accurately, which we know little about from their chemical structures and known interactions. They learn from large databases to detect patterns and correlations that may be overlooked by traditional methods. With that capability of predicting the biological activities associated with compounds, AI is purported to be able to speed up hit identification by identifying possible drug candidates for further investigation [65]. AI is also very helpful for molecular design optimization using generative models like VAEs and GANs [66]. Using thousands of different neural networks to predict human toxicity and synthesizability, these models are capable of generating new compounds that possess desired properties with no effect on efficiency or harm. The potential AI added is not only to predict activity profiles but also to suggest modifications in the hit fragments that may favor an effective medicine, making compound optimization faster and allowing quicker new therapy development [67]. The integration of AI in HCS streamlines the analysis of complex imaging data, which helps to detect even small compound-induced phenotypic changes. Through the automation of image analysis and data interpretation, AI increases both throughput and reliability in the face of HCS experiments, allowing scientists to explore more biological phenomena (and candidate drug targets thereof) [68].

Ligand-Based and Structure-Based Virtual Screening

Ligand-Based Virtual Screening

Ligand-Based Virtual Screening (LBVS) is a method in computational chemistry that finds novel drug candidates. The thinking behind this method is that molecules with similar structures are likely to have a similar function in biology. In LBVS, new or even virtual compounds are compared against a database of known active ligands to predict their potential biological activity [69]. At the heart of drug discovery, LBVS aims to expedite hit identification and lead optimization processes through prior ligand structure-activity relationships. Be it finding anticancer agents like Imatinib - a kinase inhibitor used in the treatment of chronic myeloid leukemia [70] or finding novel antiviral drugs like Remdesivir - an antiviral drug initially developed for Ebola, later repurposed for COVID-19 treatment [71]. This enabled a massive but focused screening to be carried out to sift through the possible chemical space at a relatively minimal cost and time compared to traditional experimental work. Fundamentally, Ligand-Based Virtual Screening relies on two principles - Ligand Similarity Principle and Structure Activity Relationships (SAR). The ligand similarity principle is built on the concept that compounds with similar structures tend to have similar biological activities. This fact is based on the observation that the shape, electronic configuration, and overall molecular framework have a strong influence on their interactions with biological target molecules [72]. Structure Activity Relationships (SAR) is the study of the influence of the chemical structure of a molecule on its biological activity. The aim of SAR analysis is to identify the structural features of a molecule that are responsible for its effect on biological systems.

Methods and Techniques in Ligand-Based Virtual Screening

Chemical descriptors and fingerprints are numerical representations used in molecular analysis. Descriptors encompass physical and structural properties like molecular weight, log P, and ring counts, aiding in characterizing molecules [73]. Fingerprints, on the other hand, are binary vectors indicating substructure presence, facilitating rapid similarity searches among compounds [74]. Ligand-based virtual screening employs QSAR models to correlate molecular structures with biological activities, predicting new compound activities from known examples. This approach involves cross-validation to validate model accuracy. Pharmacophore modeling identifies essential molecular features and their spatial arrangement for binding to biological targets. Molecular similarity searches gauge compound likeness using metrics such as the Tanimoto Coefficient, Euclidean Distance, or Cosine Similarity, which are crucial for drug discovery and development processes [75]. Ligand-Based Virtual Screening (LBVS) relies on extensive ligand databases like ChEMBL and PubChem, which consolidate chemical information and biological activities of known compounds, aiding hit discovery. Software tools such as Molecular Operating Environment (MOE) [76], Schrodinger's Phase, and OpenEye ROCS enhance LBVS by enabling 2D/3D similarity searches, pharmacophore modeling, and candidate molecule ranking. These tools streamline the virtual screening process, accelerating drug discovery by identifying potential lead compounds efficiently.

Workflow of LBVS

The first stage in Ligand-Based Virtual Screening (LBVS) is data pre-processing, which includes constructing a clean dataset with positive and negative examples of ligands. This is then followed by feature extraction involving the translation of the chemical structures into tuples of numbers. These are used as "descriptors," which form the basis on which predictive models such as Support Vector Machines (SVM), Random Forests, Neural Networks, etc. [77]. They are models that have been trained iteratively to discriminate between active and inactive ligands. A predictive model is trained to screen compounds using virtual screening of large compound libraries and scoring the target compound pairs based on their predicted activity. This is a high-throughput version of the process to help filter our leads. After this process, hits are identified and screened based on scoring values, binding affinity, drug-likeness, and ADMET properties [78]. These highest threshold hits are the compounds with which to proceed in drug discovery, and so efforts should be made to maximize success by focusing on these most promising initial leads.

Structure-Based Virtual Screening

Structure-Based Virtual Screening (SBVS), an alternative computational tool, can predict new ligands that bind to particular protein targets involved in the drug discovery process while omitting the huge library calculations. SBVS identifies compounds with high binding strength by analyzing the 3D structure of target proteins and known binding interactions [79]. It speeds up the time of hit identification from compound libraries, as well as reduces costs compared to traditional experimental screening approaches. Notable examples of SBVS are application in the discovery of inhibitors for HIV-1 protease, a critical enzyme in the life cycle of the HIV virus [80], and identifying inhibitors for the enzyme JAK2, which is implicated in various cancers and inflammatory diseases [81].

Principles and Concepts

Molecular docking is essential in structure-based drug design for predicting the correct orientation of a ligand with respect to its target protein [82]. This explains the conformation of bindings responsible for the biological actions taking place at protein active sites. In SBVS, molecular docking constitutes a widely used approach to predict the most likely binding modes that ligands can adopt inside protein targets, providing valuable insight for developing and refining drugs.

In molecular docking, the most important aspect is binding affinity, which defines how tightly a ligand binds to its target protein and, therefore, plays an essential role in drug discovery [83]. This is evaluated using scoring functions that compute free energy changes, including contributions such as hydrogen bonds, hydrophobic interactions, and van der Waals forces. Such interactions are at the core of accurate scoring models, which help to predict and rank candidate compounds not only in terms of potential binding but also biological activity [84].

Workflow of SBVS

SBVS begins with the target protein, a biologically relevant 3D three-dimensional structure of target proteins taken from X-ray crystallography or NMR spectroscopy. Retained structures specify the key functionalities of an active site crucial for accurate ligand orientation and screening [85]. A diverse library of small molecules and fragments is curated, which is suitable for molecular docking using in-silico design to obtain high-affinity binders. Molecular docking, the core of the SBVS workflow, involves using computational programs to predict how ligands from a library bind to a protein target's active site. Algorithms explore various binding poses based on factors like hydrogen bonds and hydrophobic interactions, assessing each ligand's fit and interaction potential [86]. Scoring functions then estimate binding affinities, helping rank ligands by their likelihood to bind strongly. Post-docking analysis validates top-ranked ligands for binding accuracy, interactions, and practical considerations such as solubility. Molecular dynamics simulations further probe complex stability, complemented by in vitro assays for experimental validation. This integrated approach optimizes drug discovery by prioritizing compounds with the highest potential for further development [87].

Tools and Software

Docking programs such as AutoDock, Schrödinger Glide, and MOE are very important in drug discovery and in determining how potential drug molecules bind to biological targets. They use complex algorithms to model and perfectly lay out the binding sites of ligand compounds for researchers so that they can choose them as potential drug candidates from a pool of compounds with millions. By predicting the binding affinities and interactions that are imperative for drug efficacy, they enable virtual screening. Docking software utilizes analysis and scoring functions like ChemScore, GoldScore, and GlideScope, which rate binding poses by interaction energy between macromolecules and geometric fitting in protein pockets [88]. These assessments are beneficial for designing drug leads that model good chances of binding with receptors and can facilitate directories like ZINC, PubChem, and DrugBank, which help in studying SARs (Structure-Activity Relationship), aiding the identification of new therapeutics.

Advantages of LBVS and SBVS

Ligand-Based Virtual Screening (LBVS) and Structure-Based Virtual Screening (SBVS) are complementary approaches in a drug discovery program. Since another important feature of LBVS is the capacity to quickly screen large compound libraries for high-throughput, this technique offers huge potential in terms of efficiency and speed. This method is not only convenient but also low-cost in terms of both time and money compared with expensive experimental assays. LBVS can be used without detailed structural knowledge of the target protein and utilize molecular descriptions (so-called chemical descriptors or fingerprints) for hit identification [89]. In addition, LBVS shows widespread utility on different biological targets and compound libraries, hence making it a versatile tool for early-phase drug discovery. On the other hand, SBVS uses structural information of the target protein to deduce ligand binding at high resolution, thereby providing more detailed mechanistic data on how a molecule binds within its catalytic site; however, nearly always with higher demand assays in terms of computational resources and expertise as it often requires expert skills in both aspects [90]. The combination of both strategies is synergistic; on one side, LBVS provides speed and versatility in comparison to SBVS, which offers a high-resolution structural detail, increasing efficiency through different stages within drug discovery pipelines.

Molecular Docking and Machine Learning Integration

Introduction

Molecular docking is a crucial aid in structural biology and drug discovery, where the goal is to predict how small molecules (ligands) bind with receptors of known or unknown 3-D structures such that an optimal value for some biological readout associated with ligand rejection can be predicted. Scoring functions to estimate the energy of an interaction and a score based on its geometrical properties, conformational search algorithms where ligand poses are evaluated (docked) binding affinity/energy calculations [91]. These principles guide the understanding and optimization of molecular interactions critical for drug design.

Machine Learning Techniques Used in Molecular Docking

Machine learning uses algorithms that learn from the repetition of operations on data to perform predictions or decisions without being explicitly programmed for them: it is known as pattern recognition and predictive modeling within the standard set of codes used in various domains. In the case of molecular docking, supervised learning is often used to predict a binding affinity or classify molecules based on training data (ligand-receptor complexes). This models the relationship between molecular properties (and, by extension, disease phenotype) and binding outcomes with input features based on molecular descriptors [92]. This method significantly improves the accuracy of virtual screening in terms of dividing compounds into binders or non-binders, thus facilitating further development with a much more focused set. Unsupervised learning in molecular docking involves analyzing unlabeled data to uncover inherent patterns. It helps to navigate chemical space by finding molecules that are bioisosteres because they have similar structures or binding modes, making the discovery of diverse drug candidates easier [93]. Furthermore, with the use of algorithms such as clustering, ligand conformations can be grouped in terms of their binding orientations or interaction patterns with receptors, which give information accompanied by found modes rather than predefined labels/outcomes. Within this process, deep learning based on multi-layered neural networks has become one of the techniques that involve studying complex molecular structures to predict different binding poses in molecular docking. These models improve scoring functions by learning from large datasets, which increases accuracy in predicting binding affinity. This application takes advantage of deep neural networks to capture complex patterns in data and represents a groundbreaking advance for computational drug discovery and molecular modeling [94].

Applications of ML in Molecular Docking

Machine learning improves molecular docking scoring functions on protein-ligand complex data. However, these traditional ways are generally not very precisely grounded in physics. Specifically, we train ML models such as regression (or even neural networks) by inputting molecular structures and empirical binding data to predict quantitative values of binding energies [95]. This improves the lead compound selection in drug discovery by capturing refined ligand-receptor interactions beyond only traditional scoring methods. Prediction of binding affinities is done by analyzing protein-ligand complexes, using machine learning to project structural and chemical information. ML models trained with molecular structures and experimental binding affinities can model these delicate interactions, such as hydrogen bonds, electrostatic forces, etc. [96]. This dramatically improves the drug development process because it gives a very accurate measure of how tightly molecules bind together, which is essential for selecting good target drugs. In conformational sampling in molecular docking, the conformational space is searched for ligands and their orientations to find optimal binding configurations. This is estimated to be covered in one shot, but there is also prediction binding poses with machine-learning accelerators that direct simulation toward high-affinity conformations [97]. This strategy improves the robustness and efficiency of this approach, meaning fewer searches must be done before suitable drug candidates are identified with high confidence while conserving computational resources.

Case Study - Synergistic Application of Molecular Docking and Machine Learning for Improved Binding Pose Prediction [98]

In this paper titled 'Synergistic application of molecular docking and machine learning for improved binding pose prediction' published in the NSO Journal, Zhang et al. explore a new approach to enhance protein-ligand complex structure predictions. This method integrates traditional molecular docking as well as more modern machine learning techniques together to attempt to overcome the weaknesses of each individual process. It is indispensable for structure-based drug design (SBDD) to effectively predict the binding modes of protein-ligand complexes. Conformational sampling algorithms and empirical scoring functions are the common properties found in traditional molecular docking, such as Autodock Vina, Uni-Dock, LeDock, etc., which have their own limitations. These techniques face challenges in the accurate description of many types of protein-ligand interactions and are limited to coverage across chemical space. However, machine learning-based scoring functions exemplified by GNINA and RTMScore have demonstrated improvements in docking success rates, albeit mainly used for re-scoring and ranking the binding poses obtained from molecular dockings. Although deep learning models such as DeepDock, KarmaDock, and Uni-Mol Docking aim to predict protein-ligand complex structures directly, the generated conformations may be incorrect because these methods are not designed to satisfy physical constraints. The workflow involved predicting binding poses utilizing deep learning models like DeepDock, KarmaDock, and Uni-Mol Docking to predict protein-ligand binding poses. Predicted binding poses are generated by the extraction of amino acid residues within a range of 10 Å of the ligand and by the regeneration of the ligand's 3D conformation using the RDKit. Position-restricted (PR) docking was performed using Uni-Dock on the predicted binding poses. This generated physically constrained and valid binding poses since the docking process focused only on the binding region. Machine learning scoring functions were applied to re-score the docked binding poses. Traditional and machine learning scoring functions like GNINA and RTMScore were used to rank the conformations and select the highest-ranked binding pose. The combined approach improved the success rate and the accuracy of the protein-ligand complex structure compared to using molecular docking and machine learning methods alone. This strategy can effectively predict protein-ligand binding poses by leveraging promises of molecular docking and machine learning methods, making it a very promising tool for rational drug design.

Strategic Integration of ML into Molecular Docking

Integrating Machine Learning and Molecular Docking Algorithms - One of the most common problems to solve is how machine learning approaches have been used in order to improve scoring functions and virtual screening processes. Machine learning algorithms identify patterns in the sets of known behaviors that would otherwise be invisible; these refine scoring to enhance prediction abilities toward binding affinity. They also enhance virtual screening by preferentially selecting binding compounds, thereby speeding drug discovery [99]. Furthermore, machine learning helps during conformational sampling to reliably place ligands in binding sites as well as substrate docking predictions. Physics-based principles are mixed with machine learning, such as in "hybrid approaches" to molecular docking, which use training sets derived from physics-informed techniques (molecular dynamics or quantum mechanics). Predictive performance is improved with ensemble methods [100]. Some of the challenges include data quality requirements, interpretability issues regarding complex ML models for validation, computational efficiency, and consideration of tool compatibility in drug discovery workflows.

Future Scope and Challenges

The future of molecular docking integrates machine learning for enhanced drug discovery, promising improved accuracy, efficiency, and personalized medicine. Deep Learning and Neural Networks may automate feature extraction, scoring functions, and conformational sampling but at the cost of data scarcity (black box effect) [101]. These involve ethical issues such as data privacy, bias correction/adaptation to regulation, and equitable access to novel therapeutics. This will require researchers, clinicians, and policy-makers to work collaboratively so as not to be unable to navigate these without succumbing honestly to the murky area of transformative biomedical applications.

Prioritization of Compound Libraries

The selection of compound libraries is a crucial process that plays a vital role in enhancing performance and productivity in the entire procedure. With these many compounds that could be synthesized and tested, there must be a way to prioritize those libraries that afford the highest probability of identifying agents capable of becoming viable drugs. Prioritization is needed in general, meaning that effective strategies are necessary for integrating biology and chemistry so as to navigate chemical space efficiently toward compounds with the biological activity of interest while maintaining flexibility and minimizing risk. The early focus on the most promising candidates enables researchers to save time, money, and effort while improving their chances of finding treatments [102].

In the drug discovery process, diversity-based strategies in compound library design are critical for increasing chances to identify more scaffolds and hit from a lead generation. This approach is knowledge-based, with structural diversity being at the core of its performance. Structural diversity in a compound library is comprised of a number of different chemical frameworks and functional groups. To inject diversity, we sample the compounds for different ring systems, functional groups, and stereochemistry. Broader coverage of chemical space in the library, therefore, increases chances for multiple scaffolds that interact with the target in distinct manners, hence serving as molecular building blocks from which lead molecules can be developed and optimized [103]. This diversity is not only great for initial hit identification but also rewarding in terms of reducing the chance of attrition further down during drug development.

In order to maintain structural diversity in a compound library, clustering techniques are used amongst other methods, such as scaffold analysis and property-based filtering [104]. Collectively, these methods have the potential to increase efficiency and reduce drug discovery failure rate by identifying novel compounds with diverse biological activities. A variation of partitioning clustering whereby the compounds are grouped together based on structural similarities so that representative members from each group can be selected to cover various parts of chemical space and reduce redundancy [105]. The idea is to embed most of the scaffolds/core structures in optimization libraries, with the intent of improving the chances of finding novel hits via some unique chemical backbones. Property-based filtering filters compounds against physicochemical properties such as molecular weight, lipophilicity, hydrogen bond donors/acceptors, and polar surface area, which seek to ensure that the compound set is both structurally diverse and drug-like.

Structural diversity in a compound library leads to different scaffolds exploring chemical space, allowing for an increased chance of finding novel hits. From this point of view, similarity-guided ways improve working library design by focusing on known active compounds with or without hitting particular protein families or chemotypes. The similarity-guided methodologies are for designing compound libraries that include structurally close compounds to the established active ones, thus focusing on targeting some specific protein families or chemotypes [106]. These insights should increase the probability of discovering new active substances in related chemical space by capitalizing on structurally well-featured compounds with validated biological activity. This strategy is especially advantageous for narrowing down to particular targets or pathways in which certain chemotypes are known to act.

Targeted library design has proven to be a highly effective tool for the identification of potent and selective ligands of particular target families, such as GPCRs or kinase inhibitors. GPCR-targeted library design strategies are based on the identification of active chemotypes and/or ligand-receptor interactions to inform scaffold selection. These methods comprise a ligand-based design with known GPCR active compounds and a structure-based design for available GPCR crystal structures [107]. In the case of kinase inhibitors, there are established libraries prepared in such a way as to quickly identify new inhibitors based on structural information derived from known kinase inhibitor complexes. For kinase-targeted libraries, there are two main objectives: 1) discovery of unique chemotypes to broaden intellectual property (IP) space and 2) desired selectivity profile by targeting specific kinases binding pockets. At the same time, the primary advantage of focused library design is that it enables the identification of active hits for a target class of interest and new chemotypes beyond those that are most prevalent. This allows the investigation of structure-activity relationships and potency/selectivity optimization. Targeted libraries, on the other hand, act against a subset of targets, such as GPCRs and kinases based only on active chemotypes. It followed a Structure, with the aid of standard and computational docking studies, moreover Ligand-Based (Pharmacophore) information to facilitate focused library design, which thereby reveals new lead molecules against significant drug targets with an ability targeted modulation [108].

Computational methods are used in molecular similarity assessment. One of the ways fingerprints are used is to represent a structure in an abbreviated way, indicating if (or not) certain bottlenecks or moieties can be observed. This allows the similarity between these fingerprints from different compounds to be determined [109]. Descriptors, on the other hand, are numeric values that represent some molecular properties such as hydrophobicity, electronic distribution, steric factors, etc. These descriptors also allow researchers to compare the textural similarity between compounds in a much finer-grain fashion [110].

The machine learning models also improve the similarity assessment as they are trained on massive datasets of active and inactive compounds. The models can recognize hard-to-find patterns or relationships that are unobservable through normal methods. Researchers can train machine learning algorithms on known active compounds and then use these models to predict the activity of new molecules based solely on their structure. With these predictive capabilities, compound libraries can be designed even more targeted and, so to speak, waste fewer valuable resources on focusing attention on the most interesting candidates [111].

Additionally, compound acquisition and prioritization algorithms cite, predict, and rank compounds based on their biological activity, diversity, and drug-ability to improve the efficiency and effectiveness of the drug discovery process. Compound acquisition and prioritization play a critical role in creating diverse and effective compound libraries during drug discovery. The acquisition of compounds demonstrates remarkable diversity using compound repositories along with strategic acquisition and commercial suppliers to construct libraries within multiple chemical structures [112]. In-silico screening methodologies pre-select compounds with increasingly high potential to lead to compounds. This ensures that the compound library contains multiple chemical structures to improve the likelihood of discovering new drug candidates with disabilities that are now producing chemical space. It means acquiring new compounds that drive beyond what is obvious in the present chemical space chemically; this significantly increases the diversity of the library [113]. In-silico screening is designed to facilitate compound acquisition by using computational methods in place of actual testing, largely consisting of the examination done on large compound databases. Methods such as molecular docking, pharmacophore modeling, and virtual screening models approximate the interaction of a compound with biological targets. The methods predict what compounds may bind and their binding affinities, followed by training inhibitors that are likely to have the desired biological activity [114]. This screening cascade helps in better utilization of resources by allowing the selection of some compounds for synthesis and testing based on their predicted inhibitory potential, thereby accelerating the early stages of drug discovery. After compounds are identified, it is necessary to prioritize which entities will best transition toward further development in order to have the largest chance of clinical success. Activity prediction models, which employ machine learning techniques to learn from the structural features and historical data of known active and inactive compounds, are used. These models are used to predict the bioactivity of new compounds, aiding in the selection process for both target prediction and experimental validation [115]. Diversity ranking algorithms select compounds that maximize chemical space coverage. Structural diversity can be quantitatively evaluated by techniques like clustering and scaffold analysis to find molecules with unique chemical frameworks, which are considered drug-like80. This maximizes the exploration of chemical space, in turn raising the possibility of identifying novel chemotypes with distinct pharmacological characteristics [116]. One approach to narrowing down the number of potentially active ligands while minimizing validation failure and using resources most effectively is through property-based filtering with respect to physicochemical metrics that are necessary for successful drug development [117]. The algorithms analyze properties such as solubility, stability, and lipophilicity to see if the advanced compounds that are being proposed have attributes that make them more "drug-like." This step increases the chance of finding hits that not only inhibit the target but also show ideal pharmaceutical properties for further development. Advanced multi-objective prioritization algorithms incorporate various factors beyond predicted activity, including diversity and physicochemical properties, to generate a more holistic ranking of compounds. The machines use the balanced relationship between these factors to help researchers create a library of compounds that is small, as well as diverse, but at the same time likely to yield more successful results when it comes to drug discovery and development [118]. New screening paradigms have emerged in drug discovery to assist research teams in reinventing the way high-throughput screening (HTS) campaigns are conducted in an effort to yield more effectiveness and increase the likelihood of success, enabling them to detect novel therapeutic candidates. In aggregate, these new experimental designs and computational methods represent a major transition to focused, data-driven strategies [119]. Computational methods are facilitating a significantly better design of HTS campaigns as well. Big data analytics, machine learning (ML), and artificial intelligence (AI) can be applied to large datasets produced by HTS screenings using high-performance computing systems for mining hidden patterns and correlations. The data-driven approaches have great potential in allowing the prioritization of compounds for synthesis based on predicted bioactivity, structural features, and drug-likeness criteria [120]. The integration of computational understanding with traditional knowledge in medicinal chemistry advances the design of compound libraries, enabling further selection for diverse chemical entities that are both bioactive and biologically relevant. These new technological approaches thus provide an engine for innovation when compared to established paradigms driven by principles., which is expected to drive substantial progress in pharmaceutical research as they offer innovative tools capable of supporting the discovery process from more secure or better treatments against a wide range of diseases. Meta-analysis is also a useful option in drug discovery to filter libraries of compounds based on data from multiple studies to obtain more solid conclusions about compound efficacy and safety. It begins with a systematic literature review of global scientific journal publications, patent records, and proprietary databases to collect all relevant studies that demonstrate statistically significant high-quality data that meet well-defined inclusion criteria. Valuable information that is of high quality, e.g., compound structures and biological activity measurements and assay conditions, are carefully extracted [121]. For each compound, effect sizes are computed via standard statistical methods, variability across studies is assessed by means of Cochran's Q or I^2 statistics, and models (fixed-effects vs. random effects) are selected based on the extent of heterogeneity. These effect sizes are combined for each compound to estimate overall efficacy, and compounds are ranked based on these estimates [122]. The subgroup analysis determines the source of heterogeneity, and sensitivity analyses are used to verify robust concreteness. Results for a given compound are then aggregated across all experiments to create an ordered list of compounds and their effect sizes on each experiment in the form of some measure (ranked list), ensuring that top-ranked ones have higher therapeutic potential than lower-rank drugs. It uses a consistent system to rank compound libraries, which increases the reliability of rankings and forces decisions supported by several studies, as stated in the meta-analysis. The process of combining these diverse and intelligent strategies not only shortens the drug discovery pipeline but also streamlines this aspect for optimized resource allocation, producing significant time as well as cost savings. The combined runs the risk of really paying off when it comes to finding at least one game-changing therapy, catalyzing a major new chapter in drug discovery that will bring dozens or scores of more effective treatments for a variety of diseases.

References

[1] Reddy, A. S., Pati, S. P., Kumar, P. P., Pradeep, H. N., & Sastry, G. N. (2007). Virtual screening in drug discovery -- a computational perspective. *Current protein & peptide science*, 8(4), 329–351. https://doi.org/10.2174/138920307781369427

[2] Martin, Y. C., Abagyan, R., Ferenczy, G. G., Gillet, V. J., Oprea, T. I., Ulander, J., ... & Zefirov, N. S. (2016). Glossary of terms used in computational drug design, part II (IUPAC Recommendations 2015). *Pure and Applied Chemistry*, *88*(3), 239-264. doi:10.1515/pac-2012-1204.

[3] Zhang, G., Guo, S., Cui, H., & Qi, J. (2018). Virtual screening of small molecular inhibitors against DprE1. *Molecules (Basel, Switzerland)*, *23*(3). doi:10.3390/molecules23030524.

[4] Hamza, A., Wei, N.-N., & Zhan, C.-G. (2012). Ligand-Based Virtual Screening Approach Using a New Scoring Function. In *Journal of Chemical Information and Modeling* (Vol. 52, Issue 4, pp. 963–974). American Chemical Society (ACS). https://doi.org/10.1021/ci200617d.

[5] Li, Q., & Shah, S. (2017). Structure-Based Virtual Screening. In *Protein Bioinformatics* (pp. 111–124). Springer New York. https://doi.org/10.1007/978-1-4939-6783-4_5.

[6] Cummins, D. J., & Bell, M. A. (2016). Integrating everything: The molecule selection toolkit, a system for compound prioritization in drug discovery. *Journal of Medicinal Chemistry*, *59*(15), 6999–7010. doi:10.1021/acs.jmedchem.5b01338.

[7] Tripathi, T., Singh, D. B., & Tripathi, T. (2024). Computational resources and chemoinformatics for translational health research. *Advances in Protein Chemistry and Structural Biology*, *139*, 27–55. doi:10.1016/bs.apcsb.2023.11.003.

[8] Gasteiger, J. (2016). Chemoinformatics: Achievements and Challenges, a Personal View. In *Molecules* (Vol. 21, Issue 2, p. 151). MDPI AG. https://doi.org/10.3390/molecules21020151.

[9] Entzeroth, M., Flotow, H., & Condron, P. (2009). Overview of high-throughput screening. *Current protocols in pharmacology*, *Chapter 9*, https://doi.org/10.1002/0471141755.ph0904s44.

[10] Broach, J. R., & Thorner, J. (1996). High-throughput screening for drug discovery. *Nature*, *384*(6604), 14-16.

[11] Astle TW. Recollections of Early Microplate Automation. *JALA: Journal of the Association for Laboratory Automation*. 2000;5(6):30-31. doi:10.1016/S1535-5535-04-00103-0.

[12] Mayr, L. M., & Bojanic, D. (2009). Novel trends in high-throughput screening. *Current Opinion in Pharmacology*, *9*(5), 580–588. doi:10.1016/j.coph.2009.08.004.

[13] Beneyton, T. et al. High-throughput screening of filamentous fungi using nanoliter-range droplet-based microfluidics. *Sci. Rep.* 6, 27223; doi: 10.1038/srep27223 (2016).

[14] Michael, S., Auld, D., Klumpp, C., Jadhav, A., Zheng, W., Thorne, N., Austin, C. P., Inglese, J., & Simeonov, A. (2008). A robotic platform for quantitative high-

throughput screening. *Assay and drug development technologies*, *6*(5), 637–657. https://doi.org/10.1089/adt.2008.150.

[15] Martis, E. A., Radhakrishnan, R., & Badve, R. R. (2011). High-throughput screening: the hits and leads of drug discovery-an overview. *Journal of Applied Pharmaceutical Science*, (Issue), 02-10.

[16] Gubler, H. (2016). High-Throughput Screening Data Analysis. In: Zhang, L. (eds) Nonclinical Statistics for Pharmaceutical and Biotechnology Industries. *Statistics for Biology and Health.* Springer, Cham. https://doi.org/10.1007/978-3-319-23558-5_5.

[17] Koga, H. (2019). Exhaustive SAR analysis tools for HTS hits based on intuition of medicinal chemists. *Chem-Bio Informatics Journal*, *19*(0), 19–31. doi:10.1273/cbij.19.19.

[18] Posner, B. A., Xi, H., & Mills, J. E. (2009). Enhanced HTS hit selection via a local hit rate analysis. *Journal of chemical information and modeling*, *49*(10), 2202–2210. https://doi.org/10.1021/ci900113d.

[19] Gayathiri, E., Prakash, P., Kumaravel, P., Jayaprakash, J., Ragunathan, M. G., Sankar, S., … Govindasamy, R. (2023). Computational approaches for modeling and structural design of biological systems: A comprehensive review. *Progress in Biophysics and Molecular Biology*, *185*, 17–32. doi:10.1016/j.pbiomolbio.2023.08.002.

[20] Chan H. C. S., Shan H., Dahoun T., Vogel H., Yuan S. Advancing Drug Discovery via Artificial Intelligence. *Trends Pharmacol. Sci.* 2019;40:592–604. doi: 10.1016/j.tips.2019.06.004.

[21] Inglese, J., Auld, D. S., Jadhav, A., Johnson, R. L., Simeonov, A., Yasgar, A., ... & Austin, C. P. (2007). Quantitative high-throughput screening: a titration-based approach that efficiently identifies biological activities in large chemical libraries. *Proceedings of the National Academy of Sciences*, 104(31), 13004-13009.

[22] Macarron, R., Banks, M. N., Bojanic, D., Burns, D. J., Cirovic, D. A., Garyantes, T., ... & Volker, P. M. (2011). Impact of high-throughput screening in biomedical research. *Nature reviews Drug discovery*, 10(3), 188-195.

[23] Iversen, P. W., Eastwood, B. J., Sittampalam, G. S., & Cox, K. L. (2006). A comparison of assay performance measures in screening assays: signal window, Z′ factor, and assay variability ratio. *Journal of biomolecular screening*, 11(3), 247-252.

[24] Lakowicz, J. R. (2006). *Principles of fluorescence spectroscopy*. Springer science & business media.

[25] Fan, F., & Wood, K. V. (2007). Bioluminescent assays for high-throughput screening. *Assay and Drug Development Technologies*, *5*(1), 127–136. doi:10.1089/adt.2006.053.

[26] Vogt, C., Wondergem, C. S., Weckhuysen, B. M. (2023). Ultraviolet-Visible (UV-Vis) Spectroscopy. In: Wachs, I. E., Bañares, M. A. (eds) *Springer Handbook of Advanced Catalyst Characterization.* Springer Handbooks. Springer, Cham. https://doi.org/10.1007/978-3-031-07125-6_11.

[27] Wolstenholme, R. (2021, January 5). Ultraviolet–Visible and Fluorescence Spectroscopy. *Analytical Techniques in Forensic Science*, pp. 115–143. doi:10.1002/9781119373421.ch6.

[28] Goodwin, S., McPherson, J. D., & McCombie, W. R. (2016). Coming of age: ten years of next-generation sequencing technologies. *Nature Reviews Genetics*, 17(6), 333-351.

[29] Barski, A., Cuddapah, S., Cui, K., Roh, T. Y., Schones, D. E., Wang, Z., ... & Zhao, K. (2007). High-resolution profiling of histone methylations in the human genome. *Cell*, 129(4), 823-837.

[30] Lister, R., Pelizzola, M., Dowen, R. H., Hawkins, R. D., Hon, G., Tonti-Filippini, J., ... & Ecker, J. R. (2009). Human DNA methylomes at base resolution show widespread epigenomic differences. *Nature*, 462(7271), 315-322.

[31] Macaulay, I. C., Ponting, C. P., & Voet, T. (2017). Single-cell multiomics: multiple measurements from single cells. *Trends in Genetics*, 33(2), 155-168.

[32] Hansel, C. S., Plant, D. L., Holdgate, G. A., Collier, M. J., & Plant, H. (2022). Advancing automation in high-throughput screening: Modular unguarded systems enable adaptable drug discovery. *Drug Discovery Today*, *27*(8), 2051-2056.

[33] N., A., C., S., & Che, T. (2013). Data analysis approaches in high throughput screening. In *Drug Discovery*. doi:10.5772/52508.

[34] Churko, J. M., Mantalas, G. L., Snyder, M. P., & Wu, J. C. (2013). Overview of high throughput sequencing technologies to elucidate molecular pathways in cardiovascular diseases. *Circulation research*, *112*(12), 1613–1623. https://doi.org/10.1161/CIRCRESAHA.113.300939.

[35] Abbas-Aghababazadeh, F., Li, Q., & Fridley, B. L. (2018). Comparison of normalization approaches for gene expression studies completed with high-throughput sequencing. *PloS one*, *13*(10), e0206312. https://doi.org/10.1371/journal.pone.0206312.

[36] Conesa, A., Madrigal, P., Tarazona, S., Gomez-Cabrero, D., Cervera, A., McPherson, A., ... & Mortazavi, A. (2016). *A survey of best practices for RNA-seq* data analysis. *Genome biology*, *17*, 1-19.

[37] Lema, Carolina & Varela-Ramirez, Armando & Aguilera, Renato. (2011). Differential nuclear staining assay for high-throughput screening to identify. *Current Cellular Biochemistry*. 1. 1-14.

[38] Murti, Della & Pujianto, Utomo & Wibawa, Aji & Akbar, Muhammad. (2019). *K-Nearest Neighbor (K-NN) based Missing Data Imputation*. 83-88. 10.1109/ICSITech46713.2019.8987530.

[39] Kumain, Sandeep & Singh, Maheep & Singh, Navjot & Berwal, Krishan. (2018). An efficient Gaussian Noise Reduction Technique For Noisy Images using optimized filter approach. 243-248. 10.1109/ICSCCC.2018.8703305.

[40] Fox, S., Farr-Jones, S., Sopchak, L., Boggs, A., Nicely, H. W., Khoury, R., & Biros, M. (2006). High-throughput screening: update on practices and success. *Journal of biomolecular screening*, *11*(7), 864-869.

[41] Malo, N., Hanley, J. A., Cerquozzi, S., Pelletier, J., & Nadon, R. (2006). Statistical practice in high-throughput screening data analysis. *Nature biotechnology*, 24(2), 167-175.

[42] Zhang, J. H., Chung, T. D., & Oldenburg, K. R. (1999). A simple statistical parameter for use in evaluation and validation of high throughput screening assays. *Journal of biomolecular screening*, 4(2), 67-73.

[43] Brideau, C., Gunter, B., Pikounis, B., & Liaw, A. (2003). Improved statistical methods for hit selection in high-throughput screening. *Journal of Biomolecular Screening*, *8*(6), 634–647. doi:10.1177/1087057103258285.

[44] Bickerton, G. R., Paolini, G. V., Besnard, J., Muresan, S., & Hopkins, A. L. (2012). Quantifying the chemical beauty of drugs. *Nature chemistry*, 4(2), 90-98. https://doi.org/10.1038/nchem.1243.

[45] JS. Zhao, Y. Guo, Q. Sheng, and Y. Shyr, "Advanced Heat Map and Clustering Analysis Using Heatmap3," *BioMed Res. Int., vol.* 2014, pp. 1–6, 2014, doi: 10.1155/2014/986048.

[46] Sarikaya and M. Gleicher, "Scatterplots: Tasks, Data, and Designs," *IEEE Trans. Vis. Comput. Graph.*, vol. 24, no. 1, pp.402–412, Jan. 2018, doi: 10.1109/TVCG. 2017.2744184.

[47] D. F. Williamson, "The Box Plot: A Simple Visual Method to Interpret Data," *Ann. Intern. Med.*, vol. 110, no. 11, p. 916, Jun. 1989, doi: 10.7326/0003-4819-110-11-916.

[48] Yao, R., Shen, Z., Xu, X., Ling, G., Xiang, R., Song, T., … Zhai, Y. (2024). Knowledge mapping of graph neural networks for drug discovery: a bibliometric and visualized analysis. *Frontiers in Pharmacology*, *15*, 1393415. doi:10.3389/fphar.2024.1393415.

[49] Hilmas, C. J. (2015). Riot Control Agents. In *Handbook of Toxicology of Chemical Warfare Agents* (pp. 131–150). doi:10.1016/b978-0-12-800159-2.00011-7.

[50] Sittampalam, G. S., Coussens, N. P., Brimacombe, K., Grossman, A., Arkin, M., Auld, D., ... & Weidner, J. (2016). *Assay guidance manual*. Eli Lilly & Company and the National Center for Advancing Translational Sciences.

[51] Motulsky, H. J., & Christopoulos, A. (2004). *Fitting models to biological data using linear and nonlinear regression: a practical guide to curve fitting*. Oxford University Press.

[52] Sebaugh, J. L. (2011). Guidelines for accurate EC50/IC50 estimation. *Pharmaceutical statistics*, 10(2), 128-134. https://doi.org/10.1002/pst.426.

[53] Sliwoski, G., Kothiwale, S., Meiler, J., & Lowe, E. W. (2014). Computational methods in drug discovery. *Pharmacological reviews*, 66(1), 334-395. https://doi.org/10.1124/pr.112.007336.

[54] Wishart, D. S. (2016). Emerging applications of metabolomics in drug discovery and precision medicine. *Nature Reviews Drug Discovery*, 15(7), 473-484. https://doi.org/10.1038/nrd.2016.32.

[55] Gaulton, A., Hersey, A., Nowotka, M., Bento, A. P., Chambers, J., Mendez, D., ... & Overington, J. P. (2017). The ChEMBL database in 2017. *Nucleic acids research*, 45(D1), D945-D954. https://doi.org/10.1093/nar/gkw1074.

[56] Bender, A., & Glen, R. C. (2004). Molecular similarity: a key technique in molecular informatics. *Organic & biomolecular chemistry*, 2(22), 3204-3218. .https://doi.org/10.1039/b409813g.

[57] Bro, R., & Smilde, A. K. (2014). Principal component analysis. *Analytical Methods*, 6(9), 2812-2831. https://doi.org/10.1039/C3AY41907J.

[58] Ekins, S., Mestres, J., & Testa, B. (2007). In silico pharmacology for drug discovery: methods for virtual ligand screening and profiling. *British journal of pharmacology*, 152(1), 9-20. https://doi.org/10.1038/sj.bjp.0707305.

[59] Schneider, G. (2018). Automating drug discovery. *Nature Reviews Drug Discovery,* 17(2), 97-113. https://doi.org/10.1038/nrd.2017.232.

[60] Svetnik, V., Liaw, A., Tong, C., Culberson, J. C., Sheridan, R. P., & Feuston, B. P. (2003). Random forest: a classification and regression tool for compound classification and QSAR modeling. *Journal of chemical information and computer sciences,* 43(6), 1947-1958. https://doi.org/10.1021/ci034160g.

[61] Burbidge, R., Trotter, M., Buxton, B., & Holden, S. (2001). Drug design by machine learning: support vector machines for pharmaceutical data analysis. *Computers & chemistry*, 26(1), 5-14. https://doi.org/10.1016/S0097-8485(01) 00094-8.

[62] Ramsundar, B., Kearnes, S., Riley, P., Webster, D., Konerding, D., & Pande, V. (2015). Massively multitask networks for drug discovery. arXiv preprint arXiv:1502.02072.

[63] Stokes, J. M., Yang, K., Swanson, K., Jin, W., Cubillos-Ruiz, A., Donghia, N. M., ... & Collins, J. J. (2020). A deep learning approach to antibiotic discovery. *Cell,* 180(4), 688-702. https://doi.org/10.1016/j.cell.2020.01.021.

[64] Menden, M. P., Wang, D., Bender, A., Augustin, M., Schüller, A., Langedijk, J., ... & Garnett, M. J. (2019). Community assessment to advance computational prediction of cancer drug combinations in a pharmacogenomic screen. *Nature communications,* 10(1), 1-17. https://doi.org/10.1038/s41467-019-09799-2.

[65] Vamathevan, J., Clark, D., Czodrowski, P., Dunham, I., Ferran, E., Lee, G., ... & Zhao, S. (2019). Applications of machine learning in drug discovery and development. *Nature Reviews Drug Discovery*, 18(6), 463-477. https://doi.org/ 10.1038/s41573-019-0024-5.

[66] Elton, D. C., Boukouvalas, Z., Fuge, M. D., & Chung, P. W. (2019). Deep learning for molecular design—a review of the state of the art. *Molecular Systems Design & Engineering,* 4(4), 828-849. https://doi.org/10.1039/C9ME00039A.

[67] Zhavoronkov, A., Ivanenkov, Y. A., Aliper, A., Veselov, M. S., Aladinskiy, V. A., Aladinskaya, A. V., ... & Zhebrak, A. (2019). Deep learning enables rapid identification of potent DDR1 kinase inhibitors. *Nature biotechnology*, 37(9), 1038-1040. https://doi.org/10.1038/s41587-019-0224-x.

[68] Caicedo, J. C., Cooper, S., Heigwer, F., Warchal, S., Qiu, P., Molnar, C., ... & Carpenter, A. E. (2017). Data-analysis strategies for image-based cell profiling. *Nature methods,* 14(9), 849-863. https://doi.org/10.1038/nmeth.4397.

[69] Oliveira, T., Silva, M., Maia, E., Silva, A., & Taranto, A. (2023). Virtual Screening Algorithms in Drug Discovery: A Review Focused on Machine and Deep Learning Methods. In *Drugs and Drug Candidates* (Vol. 2, Issue 2, pp. 311–334). MDPI AG. https://doi.org/10.3390/ddc2020017.

[70] Deininger, M., Buchdunger, E., & Druker, B. J. (2005). The development of imatinib as a therapeutic agent for chronic myeloid leukemia. In *Blood* (Vol. 105,

Issue 7, pp. 2640–2653). *American Society of Hematology*. https://doi.org/10.1182/blood-2004-08-3097.

[71] Said, M. A., Albohy, A., Abdelrahman, M. A., & Ibarhim, H. S. (2022). Remdesivir analog as SARS-CoV-2 polymerase inhibitor: virtual screening of a database generated by scaffold replacement. In *RSC Advances* (Vol. 12, Issue 35, pp. 22448–22457). Royal Society of Chemistry (RSC). https://doi.org/10.1039/d2ra00486k

[72] Wermuth, C. G., Villoutreix, B., Grisoni, S., Olivier, A., & Rocher, J.-P. (2015). Strategies in the Search for New Lead Compounds or Original Working Hypotheses. In *The Practice of Medicinal Chemistry* (pp. 73–99). Elsevier. https://doi.org/10.1016/b978-0-12-417205-0.00004-3.

[73] Todeschini, R., Consonni, V., & Gramatica, P. (2009). Chemometrics in QSAR. In *Comprehensive Chemometrics* (pp. 129–172). Elsevier. https://doi.org/10.1016/b978-044452701-1.00007-7.

[74] Yang, J., Cai, Y., Zhao, K., Xie, H., & Chen, X. (2022). Concepts and applications of chemical fingerprint for hit and lead screening. In *Drug Discovery Today* (Vol. 27, Issue 11, p. 103356). Elsevier BV. https://doi.org/10.1016/j.drudis.2022.103356.

[75] Lopez Perez, K., Avellaneda Tamayo, J., Chen, L., Lopez Lopez, E., Juarez Mercado, K. E., Medina Franco, J. L., & Miranda-Quintana, R. (2023). *Molecular Similarity: Theory, Applications, and Perspectives*. American Chemical Society (ACS). https://doi.org/10.26434/chemrxiv-2023-cs3wb.

[76] Chua, H. M., Moshawih, S., Kifli, N., Goh, H. P., & Ming, L. C. (2024). Insights into the computer-aided drug design and discovery based on anthraquinone scaffold for cancer treatment: A systematic review. In D. H. Tjahjono (Ed.), *PLOS ONE* (Vol. 19, Issue 5, p. e0301396). Public Library of Science (PLoS). https://doi.org/10.1371/journal.pone.0301396.

[77] Statnikov, A., Wang, L., & Aliferis, C. F. (2008). A comprehensive comparison of random forests and support vector machines for microarray-based cancer classification. In *BMC Bioinformatics* (Vol. 9, Issue 1). Springer Science and Business Media LLC. https://doi.org/10.1186/1471-2105-9-319.

[78] Guan, L., Yang, H., Cai, Y., Sun, L., Di, P., Li, W., Liu, G., & Tang, Y. (2019). ADMET-score – a comprehensive scoring function for evaluation of chemical drug-likeness. In *MedChemComm* (Vol. 10, Issue 1, pp. 148–157). Royal Society of Chemistry (RSC). https://doi.org/10.1039/c8md00472b.

[79] Lionta, E., Spyrou, G., Vassilatis, D., & Cournia, Z. (2014). Structure-Based Virtual Screening for Drug Discovery: Principles, Applications and Recent Advances. In *Current Topics in Medicinal Chemistry* (Vol. 14, Issue 16, pp. 1923–1938). Bentham Science Publishers Ltd. https://doi.org/10.2174/1568026614666140929124445.

[80] Weber, I. T., Wang, Y.-F., & Harrison, R. W. (2021). HIV Protease: Historical Perspective and Current Research. In *Viruses* (Vol. 13, Issue 5, p. 839). MDPI AG. https://doi.org/10.3390/v13050839.

[81] Lin, T. E., HuangFu, W. C., Chao, M. W., Sung, T. Y., Chang, C. D., Chen, Y. Y., Hsieh, J. H., Tu, H. J., Huang, H. L., Pan, S. L., & Hsu, K. C. (2018). A Novel

Selective JAK2 Inhibitor Identified Using Pharmacological Interactions. In *Frontiers in Pharmacology* (Vol. 9). Frontiers Media SA. https://doi.org/10.3389/fphar.2018.01379.

[82] Torres, P. H. M., Sodero, A. C. R., Jofily, P., & Silva-Jr, F. P. (2019). Key Topics in Molecular Docking for Drug Design. In *International Journal of Molecular Sciences* (Vol. 20, Issue 18, p. 4574). MDPI AG. https://doi.org/10.3390/ijms 20184574.

[83] Agu, P. C., Afiukwa, C. A., Orji, O. U., Ezeh, E. M., Ofoke, I. H., Ogbu, C. O., Ugwuja, E. I., & Aja, P. M. (2023). Molecular docking as a tool for the discovery of molecular targets of nutraceuticals in diseases management. In *Scientific Reports* (Vol. 13, Issue 1). Springer Science and Business Media LLC. https://doi.org/10.1038/s41598-023-40160-2.

[84] Velasquez-López, Y., Tejera, E., & Perez-Castillo, Y. (2022). Can docking scoring functions guarantee success in virtual screening? In *Virtual Screening and Drug Docking* (pp. 1–41). Elsevier. https://doi.org/10.1016/bs.armc.2022.08.008.

[85] Waszkowycz, B., Clark, D. E., & Gancia, E. (2011). Outstanding challenges in protein–ligand docking and structure-based virtual screening. In *WIREs Computational Molecular Science* (Vol. 1, Issue 2, pp. 229–259). Wiley. https://doi.org/10.1002/wcms.18.

[86] Dias, R., & de Azevedo Jr., W. (2008). Molecular Docking Algorithms. In *Current Drug Targets* (Vol. 9, Issue 12, pp. 1040–1047). Bentham Science Publishers Ltd. https://doi.org/10.2174/138945008786949432.

[87] Salo-Ahen, O. M. H., Alanko, I., Bhadane, R., Bonvin, A. M. J. J., Honorato, R. V., Hossain, S., Juffer, A. H., Kabedev, A., Lahtela-Kakkonen, M., Larsen, A. S., Lescrinier, E., Marimuthu, P., Mirza, M. U., Mustafa, G., Nunes-Alves, A., Pantsar, T., Saadabadi, A., Singaravelu, K., & Vanmeert, M. (2020). Molecular Dynamics Simulations in Drug Discovery and Pharmaceutical Development. In *Processes* (Vol. 9, Issue 1, p. 71). MDPI AG. https://doi.org/10.3390/pr9010071.

[88] Yang, C., Chen, E. A., & Zhang, Y. (2022). Protein–Ligand Docking in the Machine-Learning Era. In *Molecules* (Vol. 27, Issue 14, p. 4568). MDPI AG. https://doi.org/10.3390/molecules27144568.

[89] Geppert, H., Vogt, M., & Bajorath, J. (2010). Current Trends in Ligand-Based Virtual Screening: Molecular Representations, Data Mining Methods, New Application Areas, and Performance Evaluation. In *Journal of Chemical Information and Modeling* (Vol. 50, Issue 2, pp. 205–216). American Chemical Society (ACS). https://doi.org/10.1021/ci900419k.

[90] Ferreira, L., Dos Santos, R., Oliva, G., & Andricopulo, A. (2015). Molecular Docking and Structure-Based Drug Design Strategies. In *Molecule*s (Vol. 20, Issue 7, pp. 13384–13421). MDPI AG. https://doi.org/10.3390/molecules200713384.

[91] Meli, R., Morris, G. M., & Biggin, P. C. (2022). Scoring Functions for Protein-Ligand Binding Affinity Prediction Using Structure-based Deep Learning: A Review. In *Frontiers in Bioinformatics* (Vol. 2). Frontiers Media SA. https://doi.org/10.3389/fbinf.2022.885983.

[92] Le Menestrel, Thomas & Rivas, Manuel. (2024). Smiles2Dock: an open large-scale multi-task dataset for ML-based molecular docking.

[93] Pushkaran, A. C., & Arabi, A. A. (2024). From understanding diseases to drug design: can artificial intelligence bridge the gap? In *Artificial Intelligence Review* (Vol. 57, Issue 4). Springer Science and Business Media LLC. https://doi.org/10.1007/s10462-024-10714-5

[94] Turzo, S. B. A., Hantz, E. R., & Lindert, S. (2022). Applications of machine learning in computer-aided drug discovery. In *QRB Discovery* (Vol. 3). Cambridge University Press (CUP). https://doi.org/10.1017/qrd.2022.12

[95] Schapin, N., Majewski, M., Varela-Rial, A., Arroniz, C., & Fabritiis, G. D. (2023). Machine learning small molecule properties in drug discovery. In *Artificial Intelligence Chemistry* (Vol. 1, Issue 2, p. 100020). Elsevier BV. https://doi.org/10.1016/j.aichem.2023.100020.

[96] Gómez-Tamayo, J. C., Cao, L., Ahmad, M., & Tresadern, G. (2024). Binding Affinity Prediction with 3D Machine Learning: Training Data and Challenging External Testing. *American Chemical Society* (ACS). .https://doi.org/10.26434/chemrxiv-2024-s5sbs

[97] Shamim, S., Munawar, R., Rashid, Y., Muhammad Zesshan Qadar, S., Bushra, R., Begum, I., Imran, M., & Quds, T. (2024). Molecular Docking: An Insight from Drug Discovery to Drug Repurposing Approach. In *Unravelling Molecular Docking - From Theory to Practice* [Working Title]. IntechOpen. https://doi.org/10.5772/intechopen.1005526.

[98] Li, Y., Lin, H., Yang, H., Yuan, Y., Zou, R., Zhou, G., Zhang, L., & Zheng, H. (2024). Synergistic Application of Docking and Machine Learning for Improved Binding Pose. In *National Science Open.* Science China Press., Co. Ltd. https://doi.org/10.1360/nso/20230058.

[99] Han, R., Yoon, H., Kim, G., Lee, H., & Lee, Y. (2023). Revolutionizing Medicinal Chemistry: The Application of Artificial Intelligence (AI) in Early Drug Discovery. In *Pharmaceuticals* (Vol. 16, Issue 9, p. 1259). MDPI AG. https://doi.org/10.3390/ph16091259.

[100] Karniadakis, G. E., Kevrekidis, I. G., Lu, L., Perdikaris, P., Wang, S., & Yang, L. (2021). Physics-informed machine learning. In *Nature Reviews Physics* (Vol. 3, Issue 6, pp. 422–440). Springer Science and Business Media LLC. https://doi.org/10.1038/s42254-021-00314-5.

[101] Gabel, J., Desaphy, J., & Rognan, D. (2014). Beware of Machine Learning-Based Scoring Functions—On the Danger of Developing Black Boxes. In *Journal of Chemical Information and Modeling* (Vol. 54, Issue 10, pp. 2807–2815). American Chemical Society (ACS). https://doi.org/10.1021/ci500406k.

[102] Walters, W. P., Stahl, M. T., & Murcko, M. A. (1998). Virtual screening—an overview. *Drug discovery today*, *3*(4), 160-178.

[103] Lipkus, A. H., Yuan, Q., Lucas, K. A., Funk, S. A., Bartelt, W. F., 3rd, Schenck, R. J., … CAS Registry. (2008). Structural diversity of organic chemistry. A scaffold analysis of the CAS Registry. *The Journal of Organic Chemistry*, *73*(12), 4443–4451. doi:10.1021/jo8001276.

[104] Lipinski, C. A. (2004). Lead-and drug-like compounds: the rule-of-five revolution. *Drug Discovery Today: Technologies,* 1(4), 337-341. https://doi.org/10.1016/j.ddtec.2004.11.007.

[105] Besnard, J., Ruda, G. F., Setola, V., Abecassis, K., Rodriguiz, R. M., Huang, X. P., ... & Hopkins, A. L. (2012). Automated design of ligands to polypharmacological profiles. *Nature,* 492(7428), 215-220. https://doi.org/10.1038/nature11691.

[106] Schreiber, S. L. (2000). Target-oriented and diversity-oriented organic synthesis in drug discovery. *Science,* 287(5460), 1964-1969. https://doi.org/10.1126/science.287.5460.1964.

[107] Lowrie, J., Delisle, R., Hobbs, D., & Diller, D. (2004). The different strategies for designing GPCR and kinase targeted libraries. *Combinatorial Chemistry & High Throughput Screening*, *7*(5), 495–510. doi:10.2174/1386207043328625.

[108] Lipkin, M. J., Stevens, A. P., Livingstone, D. J., & Harris, C. J. (2008). How large does a compound screening collection need to be?. *Combinatorial Chemistry & High Throughput Screening*, *11*(6), 482-493.

[109] Tan, L., Lounkine, E., & Bajorath, J. (2008). Similarity searching using fingerprints of molecular fragments involved in protein− ligand interactions. *Journal of chemical information and modeling*, *48*(12), 2308-2312.

[110] Mauri, A., Consonni, V., & Todeschini, R. (2017). Molecular Descriptors. In *Handbook of Computational Chemistry* (pp. 2065–2093). doi:10.1007/978-3-319-27282-5_51.

[111] Lavecchia, A. (2015). Machine learning methods in early drug discovery. Current drug targets, 16(6), 588-604. https://doi.org/10.2174/1389450116666150202160332.

[112] Ma, C., Lazo, J. S., & Xie, X. Q. (2011). Compound acquisition and prioritization algorithm for constructing structurally diverse compound libraries. *ACS Combinatorial Science*, *13*(3), 223–231. doi:10.1021/co100033m.

[113] Medina-Franco, J. L., Giulianotti, M. A., Welmaker, G. S., & Houghten, R. A. (2013). Shifting from the single to the multitarget paradigm in drug discovery. *Drug discovery today*, 18(9-10), 495-501. https://doi.org/10.1016/j.drudis.2013.01.008.

[114] Chang, Y., Hawkins, B. A., Du, J. J., Groundwater, P. W., Hibbs, D. E., & Lai, F. (2022). A Guide to In Silico Drug Design. *Pharmaceutics*, *15*(1), 49. https://doi.org/10.3390/pharmaceutics15010049.

[115] Stokes, J. M., Yang, K., Swanson, K., Jin, W., Cubillos-Ruiz, A., Donghia, N. M., ... & Collins, J. J. (2020). A deep learning approach to antibiotic discovery. *Cell,* 180(4), 688-702. https://doi.org/10.1016/j.cell.2020.01.021.

[116] Geppert, H., Vogt, M., & Bajorath, J. (2010). Current trends in ligand-based virtual screening: molecular representations, data mining methods, new application areas, and performance evaluation. *Journal of chemical information and modeling*, 50(2), 205-216.

[117] Gaulton, A., Bellis, L. J., Bento, A. P., Chambers, J., Davies, M., Hersey, A., ... & Overington, J. P. (2012). ChEMBL: a large-scale bioactivity database for drug discovery. *Nucleic acids research*, 40(D1), D1100-D1107.

[118] Nicolaou, C. A., & Brown, N. (2013). Multi-objective optimization methods in drug design. *Drug Discovery Today: Technologies*, 10(3), e427-e435. https://doi.org/10.1016/j.ddtec.2013.02.001.

[119] Handl, J., Knowles, J., & Kell, D. B. (2005). Computational cluster validation in post-genomic data analysis. Bioinformatics, 21(15), 3201-3212. https://doi.org/10.1093/bioinformatics/bti517.

[120] Pyzer-Knapp, E. O., Pitera, J. W., Staar, P. W. J., Takeda, S., Laino, T., Sanders, D. P., … Curioni, A. (2022). Accelerating materials discovery using artificial intelligence, high performance computing and robotics. *Npj Computational Materials*, 8(1). doi:10.1038/s41524-022-00765-z.

[121] Sutton, A. J., Abrams, K. R., Jones, D. R., Sheldon, T. A., & Song, F. (2000). *Methods for meta-analysis in medical research*. John Wiley & Sons.

[122] Sacks, H. S., Berrier, J., Reitman, D., Ancona-Berk, V. A., & Chalmers, T. C. (1987). Meta-analyses of randomized controlled trials. *New England Journal of Medicine,* 316(8), 450-455.

Chapter 5

Unsupervised Learning and Dimensionality Reduction: Applications in Drug Discovery

Pooja Hoovina Venkatesh[1,*]
Sanchana Mishra[2,#]
Vanshika Bhatia[2,#]
and Vidya Niranjan[1,†]
[1]RV College of Engineering, Bengaluru, affiliated with Visvesvaraya Technological University, Belgaum, India
[2]Department of Zoology, St. Joseph's University, Karnataka, Bengaluru

Abstract

Computer-aided drug discovery (CADD) approaches effectively reduce the cost and time required for a new drug discovery. Moreover, the emergence of artificial intelligence (AI)/machine learning (ML) has revolutionized the pharmaceutical industry by providing innovative solutions to traditional drug discovery challenges.

In this chapter, we will discuss the unsupervised learning and dimensionality reduction algorithms of AI/ML, highlighting the significance of unsupervised machine learning inenabling the analysis of complex and high-dimensional unlabeled data, which is particularly useful in the early stages of drug discovery. We will also emphasize the key techniques, such as clustering, t-distributed stochastic neighbour embedding (t-SNE), and Variational Autoencoders, along with

---

* Corresponding Author's Email: vidya.n@rvce.edu.in
# These authors have contributed equally.
† Corresponding Author's Email: vidya.n@rvce.edu.in

In: Harnessing AI for Drug Discovery
Editor: Kolla Bhanu Prakash
ISBN: 979-8-89530-540-9

challenges associated with the interpretability and reproducibility of results from the unsupervised models.
In summary, this chapter highlights the potential applications of unsupervised machine learningand dimensionality reduction techniques in the field of drug discovery and development pipeline.

Keywords: unsupervised learning, principal component analysis, t-SNE, autoencoders, computer-aided drug discovery

Introduction

Overview of the Drug Discovery and Development Process

Drug discovery involves finding a chemical molecule with curative potential to treat a disease.This process of rendering a chemical molecule from lab to market requires significant time andeffort. On average, out of millions of molecules tested, only a few lead molecules are investigated in advanced clinical trials and made accessible to patients after FDA approval (Deore et al., 2019).

Therapeutic compounds, also known as pharmaceuticals, were traditionally discovered by accidental incidents, either through serendipitous discoveries like penicillin, chlorpromazine, and aniline purple or by extracting the active ingredient from a plant or flower as in a traditionalremedy (Ban, 2006). Later, with the advancement of technology, the active ingredients were synthetically reproduced in a controlled laboratory setting. For instance, morphine, a phytochemical extracted from *Papaver somniferum,* has a history of about 200 years through its therapeutic applications in analgesia and anesthesia (Katiyar et al., 2012; Wicks, Hudlicky,& Rinner, 2021).

The history of the drug discovery process in the modern era can be divided into three distinct periods of development. Therapeutic chemists and serendipitous findings in the 19th century were the basis for drug discovery throughout the first significant period. The second period ofantibiotic research was started in the early 20th century with the determination of inventive drug structures, which facilitated the development of more innovations in this field through advanced computational techniques towards the end of the century. The advent of rDNA technology and the 21st century 'Omics' revolution has revolutionized the third period of drugdiscovery (Pina, Hussain, & Roque, 2009).

The drug discovery pipeline encompasses several phases and processes, from conceptualization to development to approval. The primary steps in drug discovery and development involve target identification, high-throughput screening of compounds, optimization of lead compounds for their potency and selectivity, the conduct of extensive experimental testing, the evaluation of the safety and efficacy of potential drugs through animaland human trials, and, if successful, undergoing a regulatory review for approval of the new drug followed by post-marketing surveillance (Singh et al., 2023).

Integrating computational approaches at various stages of this process has significantly reducedthe cost and time required to discover new drugs. Several biotech and pharmaceutical companies depend on the intelligent software that involves CADD approaches. CADD is subdivided into two categories: ligand-based drug discovery (LBDD) and structure-based drug discovery (SBDD). A few examples of therapeutic drugs, including Aliskiren, Captopril, Oseltamivir, Dorzolamide, and many more, were discovered and developed using thesecomputational techniques at various stages of the drug discovery process. In recent times, advanced AI and ML algorithms, along with high performance and quantum computation, have helped resolve multiple challenges associated with drug discovery (Ece, 2023; Santagati et al.,2024).

Introduction to Artificial Intelligence/Machine Learning (AI/ML)

AI is a computer science-based field where the systems can do duties that typically require human intelligence. This encompasses a broad range of multidisciplinary technologies that enable machines to mimic human intelligence. AI comprises rule-based programming, ML, and deep learning (DL) algorithms (Georgevici & Terblanche, 2019). The main objectiveof ML and DL subtypes of AI is to automate tasks (or) duties that need human intelligence (Choi et al., 2020). These advanced methods are used in multiple fields, such as healthcare, manufacturing, logistics, and agriculture, to enhance efficiency and decisive processes. Also, various applications of AI are evident in its role in robotics, data science, extensive data management, image recognition, natural language processing, geographic information systems (GIS), nanotechnology, and cybersecurity (Ijaz et al., 2023; Yarali, 2023).

Machine learning (ML) models generate a solution (or) answer using subsets of data based on distinct combinations of critical highlights (or) patterns, but in contrast, classical programming methods are used to code a calculation using known characteristics (Badillo et al., 2020; Gupta et al., 2021). For example, deep neural network models are trained using mechanical time-frequency signals. This trained model enables automatic feature extraction and intelligent fault feature diagnosis, eliminating dependence on signal processing technology. A study proposed a SubCNN model for multiscale feature extraction and fault diagnosis with a fault recognition rate greater than 99.7% on the dataset (Yuan et al., 2022).

ML methods can be divided into two main categories: supervised (S) and unsupervised(US) learning algorithms, which are task-driven and data-driven techniques, respectively. Thisclassification of the ML algorithms is illustrated in Figure 1. Supervised learning (SL) modelsuse labeled input training data to make forecasts or recognize patterns, whereas unsupervised learning (USL) models use unlabeled input training data. Some examples of USL algorithms extensively used are principal component analysis (PCA) and hierarchical clustering (Camacho et al., 2018). The various USL algorithms are highlighted in Figure 1, as this chapter focuses on USL techniques. In a recent study, semi-supervised learning (SSL) and reinforcement learning (RL) algorithms were highlighted as two more categories of ML methods. SSL modelsare reported to use a smaller fraction of labeled datasets combined with larger unlabeled datasets during the training process. Hence, these algorithms are also known as hybridization learning algorithms. On the other hand, RL is an environmental data-driven algorithm in whichinformation gathered through activists from an environment is used to take the necessary actions for better outcomes (Sarker, 2021).

A study reported the applications of these advanced techniques at various stages of new drug identification, which helps overcome new drug discovery challenges like off-targetdelivery and low efficacy. Effectively reducing these challenges influences cost and time consumption (Gupta et al., 2021). Nowadays, most of these algorithms are being used in various steps that include in-silico synthesis of small molecules, hit identification through structure and ligand-based virtual screening, lead molecule prediction, drug dosage, molecular potency detection, quantitative structure-activity relationship (QSAR) modeling, bioactivity predictions, drug toxicity, and efficacy studies. Along with these applications, USL models arealso used in pharmacodynamic and pharmacokinetic modelling, protein-protein interaction, and protein folding studies (Gupta et al., 2021).

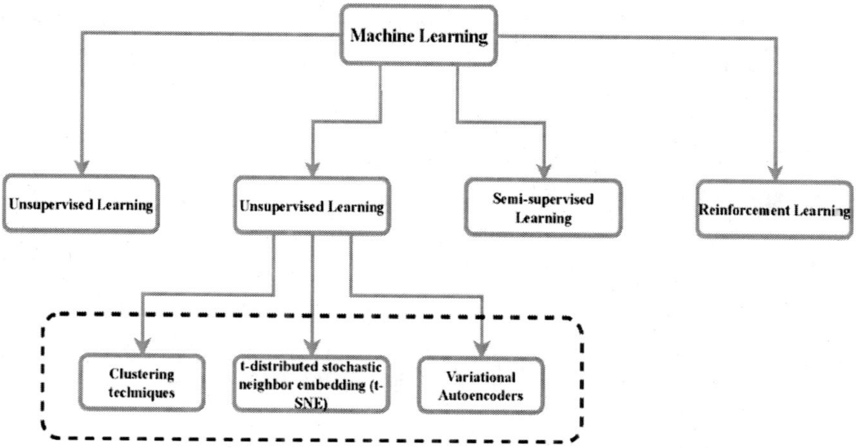

Figure 1. Classification of machine learning algorithms.

Overview and Significance of Unsupervised Learning (USL) in Drug Discovery

As mentioned in Section 4.0.2, USL algorithms use complex, unlabeled data to detect hidden patterns and structures. A few examples of such unlabeled big data include the data of pharmaceutical components and complex biological systems. This approach is beneficial when the entire dataset lacks labeled highlights, usually occurring in early drug discovery stages (Dara et al., 2022; Polanski, 2022). The conventional approach to drug discovery involves manual experimentation techniques, which include high-throughput screening and analysis andtrial-and-error experimental procedures (Burzykowski et al., 2023). However, with the arrivalof AI and ML, novel opportunities have emerged for helping both academics and industry analyze large datasets more precisely and orderly. USL approaches are being employed at several stages, such as *de-novo* molecular design, virtual screening, and efficacy and toxicity predictions of therapeutic candidates (Blanco-González et al., 2023; Chen et al., 2023). It has been extensively utilized in drug design, particularly in Computer-assisted Synthesis Design (CASD), which uses a synthesis tree to explore potential molecule acquisition methods, demonstrating its productivity and need for large-scale molecular information processing (Polanski, 2022).

A few of the applications of USL in the CADD pipeline that have been reported in a few studies are mentioned below:

- *Mapping molecular representation during drug design*: Variational autoencoders are a tool of the unsupervised learning approach employed in the automatic drug design process. This algorithm converts the SMILES to a continuous molecular representation of the candidatemolecule, which helps analyze the similarities between the known inhibitors and the candidate molecule. With this, we can swiftly and competently traverse chemical spaces toexpose feasible therapeutic candidates (Polanski, 2022).
- *De novo drug design*: Another study also reported using variational autoencoders and generative hostile networks, two high-yielding approaches to generating de novo drug candidates with desired properties (Dara et al., 2022).
- *Target identification and validation*: The same study reported that clustering algorithms could concede and validate new therapeutic targets by disclosing relationships between drug compounds, disease targets, and disorder characteristics (Dara et al., 2022).
- *Prognostic biomarker prediction*: The study also states that unsupervised algorithms can survey high-dimensional omics data to identify predictive biomarkers for medical applications (Dara et al., 2022).

New small molecular therapeutic discoveries often involve exploring complex datasets in which entire data information may not be generally labelled or measured at every stage. Insuch cases, USL models seem more appropriate than SL models. This helps understand thelarge chemical spaces and complex biological structures, leading to the discovery of noveltherapies (Bateman et al., 2016; Chen et al., 2023; Dara et al., 2022; Patil et al., 2022).

Because of this wide range of applications, it is crucial to understand the underlying concepts behind these algorithms. This knowledge about the principles of USL models would help solve more complicated challenges requiring these models. Therefore, the mainobjective of this book chapter is to understand various USL techniques algorithms by specifically discussing PCA, dimensionality reduction (DR), t-SNE algorithms, autoencoders, and clustering techniques. Here, we will also highlight the application of these USL models in pharmaceutical industries.

Principal Component Analysis (PCA)

Introduction to Principal Component Analysis

PCA is a data reduction tool used to develop hypotheses based on multidimensional statistical mechanics. It is used to model biological systems without understanding the underlying theoretical assumptions. During PCA modelling of biological systems, data reduction helps us identify hidden patterns and relationships that would otherwise make gaining insights into a particular biological process challenging. Hence, it can be a crucial technique for addressing drug discovery challenges from a broader perspective beyond conventional approaches (Giuliani, 2017).

Reducing the dataset's dimensionality could help preserve its essential features. This isachieved by its variance in a vector space. In this vector space, based on the relationship between the information retrieved in the dataset, the responses from this analysis are investigated as an outcome of the specific treatment conditions on the sample (Alizadeh Behbahani et al., 2017).

In the data analysis of electroencephalogram results, dimension reduction using PCA is usually recommended before the disintegration of electroencephalogram data by independentcomponents analysis (ICA). This approach reduces the computation time and the amount of data required (Artoni, Delorme, & Makeig, 2018). The PCA algorithm estimates the original information table with a few key elements. This primer thoroughly explains a technique, encompassing its geometry, definition, and analysis of the outcomes regarding numbers and graphics. It addresses the primary graphical output, frequently displayed as a biplot, and its variations, including grouped data analysis and correspondence analysis (Greenacre et al., 2022).

PCA involves two main components for plotting the graphs: empirical orthogonal functions (EOFs) and principal component (PC) time series ("Principal Component Analysis," 2022). However, the number of crucial components could be precisely determined according to the modelling requirements. This could be achieved using cross-validation (Wold, Esbensen, & Geladi, 1987).

PCA is a parameter-free and assumption-free statistical technique that can be used withany mathematical dataset, regardless of capacity. This involves complicated computations andcould be untraceable. The results obtained from a PCA analysis could be visualized on scatterplots with preserved vital patterns. The capabilities of PCA can be assessed by reflectingon the quantity of information that has been acknowledged to have extraordinary volatility. Because of its application on mathematical datasets, these algorithms have been incorporated into various analysis packages, such as EIGENSOFT and PLINK. There could be no unanimous approval about how many PCs need to be examined; nonetheless, it satisfies replicability risk norms and could be frequently employed in the study of population genetics and related fields (Elhaik, 2022).

Application of PCA in the CADD Process

PCA has various applications, such as calculating missing values in big data matrices, estimating sparse components, and studying shapes, functions, and pictures (Greenacre et al., 2022). It is also helpful in guiding chemical library designs. In the process of discovering new drugs, this approach has been shown to be very important as the structure and function of operating drugs fall within a narrow range (Wenderski et al., 2015).

The authors, Yoo and Shahlaei, extensively used PCA algorithms in one of their studies on a dataset of CCR5 inhibitors. They evaluated the PCA performance by comparing differentpreprocessing methods. This evidence shows the crucial application of these techniques in QSAR studies and in the analysis of data matrices where molecules are described by various interrelated molecular descriptors (Yoo & Shahlaei, 2018).

Yet another study showed the use of PCA analysis in determining fungi with multiple bioactivities. In this study, the bioactivities of fungi were effectively analyzed and filtered usingPCA analysis. This resulted in an array of data used to examine and choose multi-bioactive fungi, including *Aspergillus clavatus, Aspergillus flavipes*, and *Aspergillus citrinoterreus* (Yang et al., 2022).

PCA analysis involves exploring complex datasets and can be used in clinical settings to analyse brain injuries. This was highlighted in a study in which the PCA approach was utilized to analyse the topography of hypoxic-ischemic brain injury, contrasted with the probability method. The t-statistics for the probability of the data and determination of the angle between the PCs for covariance data were also employed during these analyses (Singhalet al., 2012). Hence, the above studies showcase the potential of PCA in identifying natural products and new drugs, which can positively impact drug development and diagnosticapplications.

The discrepancies within a dataset can undermine PCA's ability to identify the actual low- dimensional structure. This can be addressed using the PCA median of means (MoM) approach, resulting in MoMPCA. The method offers exceptional consolidation rates and lower error expectations. This approach provides novel insights on dimensions and avoids some non-asymptotic inaccuracies through Rademacher complexities in analysing data in a Hilbert space.Real-world data and simulations show that this concept is practical, as demonstrated by Paul, Chakraborty, and Das (Paul, Chakraborty, & Das, 2024).

To achieve the desired outcome, the results of PCA are easily manipulated, which is considered an artifact of PCA. So, using PCA can lead to adverse consequences in collaborativeresearch. Hence, the findings of PCA may not be robust, consistent, or reproducible under accepted standards in the field. (Elhaik, 2022).

T-Distributed Stochastic Neighbour Embedding (t-SNE)

Overview of Dimensionality Reduction and t-SNE Algorithms

Dimensionality reduction refers to the extrapolation of data from a high-dimensional space into a low-dimensional space. In this process of reducing the data from a high-dimensional space to a lower-dimensional space, similar datasets are brought close to each other to fit in a constrained space. This helps preserve the data's critical characteristics in its reduced form (Zhang et al., 2016). Thus, a reduction in the complexity of the data makes iteasier to visualize, analyze, and demonstrate complex data (Hagele, Krake, & Weiskopf, 2023).For instance, genomic data is high-dimensional, which imposes challenges during analysis due to overfitting the data, longer computation times, and giant space capacity. Suchdata can be dimensionally reduced and visualized as clusters and development trajectoriesprojected from high- onto a low-dimensional space, usually in a two- or three-dimensionalplane (Ding, Condon, & Shah, 2018; Gola et al., 2016). The study conducted by Ding et al.stated the use of DR of the transcriptome data from single-cell RNA-sequencing methodbecause of the potential ability to conserve both the local and global structural properties of thedata, thus improving its interpretability (Ding, Condon, & Shah, 2018). The uniform manifoldapproximation and projection (UMAP) algorithm is one nonlinear dimensionality-reductiontool that was developed to analyze complex high-dimensional data and was shown to be appliedto well-characterized biological data like mass cytometry and single-cell RNA sequencing datasets (Becht et al., 2018).

DR methods can also be used to analyze data from a molecular dynamic simulation ofmacromolecules and visualize this data in lower space. However, a significant loss of the structural data limits the success of the analysis. The USL methods that are commonly used forsuch analysis include PCA, pre-defined order parameters, collective variables (CVs), and time-structure-based independent component analysis (t-ICA) (Zhou, Wang, & Tao, 2018).

t-SNE is a non-linear DR algorithm used for statistically clustering the high- dimensional data and visualizing these data points in a 2- or 3-D space at specific locations (van der Maaten & Hinton, 2008). This technique was developed by Van Der Maaten and Hinton in 2008, modified from stochastic neighbor embedding, which was initially stated by Geoffrey Hinton and Sam Roweis (Taylor, Hinton, & Roweis, 2007; van der Maaten & Hinton,2008). Two main steps are involved in t-SNE. The first step is to reduce data from a high- dimensional to a low-dimensional space, including the probability distribution of data based on the similarity. The next step is to minimize the distance between two distributions by the Kullback-Leibler divergence (KL divergence) algorithm following the data positions in the low-dimensional space (Tang et al., 2024). None of the other visualization techniques, like Sammon mapping, locally linear embedding, or Isomap, are as effective as the t-SNE algorithmfor many datasets (van der Maaten & Hinton, 2008).

The Importance of t-SNE in the CADD Process

In the drug discovery process, bioinformatic tools play a significant role in analyzing the biological data and properties of drug-likeliness, which helps identify the potential drug molecules based on their properties (Prabha et al., 2023). The recent advances in genomics, transcriptomics, and proteomics because of next-generation sequencing technologies are enhancing drug development and leading to massive amounts of high-dimensional datasets (Hussain, 2019). Routinely used analysis techniques may not be feasible to handle such datasets. Therefore, integrating t-SNE algorithms with bioinformatics and sequencing data analysis might offer a holistic approach to identifying new drug molecules and targeted therapeutics.

These algorithms can be easily integrated with other computational strategies due to their feasibility in working with a multi-faceted pipeline. A study reported the integration of t-SNE analysis with ECFP4 fingerprints. It was used to analyze the molecular similarities between known clinical trial inhibitors and FDA-approved drugs. This approach helped the authors identify Palbociclib and Ribociclib analogs as potent WEE1 kinase inhibitors. Hence, this study demonstrated the potential role of the t-SNE algorithm in drug discovery for targetedcancer therapy (Muthuraj et al., 2024).

Yet another study on predicting the off-target interactions for small molecular drugs using computational approaches has stated t-SNE analysis. This particular analysis was used to derive the chemical diversity distributions of 857 drugs, including 401 clinically approved drugs (Rao et al., 2019). From these studies, we could say that t-SNE analysis helps uncover the hidden relationships and patterns in the dimensionally reduced data. And this might ultimately influence the timeline of drug innovation.

Therefore, t-SNE analysis enables us to analyze and visualize complex data in a low- dimensional space, making it an important tool for both academic and industrial environments.

Autoencoders and Variational Autoencoders

A Brief Introduction to Autoencoders

Autoencoder is a USL circuit model with a masked layer whose primary purpose is to reduce the average manipulation measure between inputs and outputs (Baldi & Lu, 2012). It isreported to be formerly called an auto-associative multilayer perceptron (Bourlard & Kabil, 2022). The autoencoders generally contain three parts: the encoder, the latent space, and the decoder (Singh & Ogunfunmi, 2021). A schematic representation of an autoencoder is shown in Figure 2.

Autoencoders (AE) are neural networks that can serve as the dimensionality reduction of data (Zhang et al., 2019) and recreate d-dimensional input vectors as d-dimensional output vectors. These extract code on the latent encoding layer and decode at the output layer, which helps minimize mean square error (Bourlard & Kabil, 2022). Linear autoencoders are used forlinear transformations between visible and hidden variables across variable fields, but only real-valued data have been extensively investigated.

Boolean autoencoders are non-linear autoencoders with Boolean functions. These could have deep connections to clustering, makingtheir comparison with linear autoencoders from a clustering perspective equitable (Baldi & Lu,2012). Apart from DR, these autoencoders could also have several applications in object recognition, image categorization, and denoising (Singh & Ogunfunmi, 2021).

Variational autoencoders (VAEs) are DL models; they acquire the knowledge required for the probability distribution of some data set and express it to a latent space. These ML models could transform the input data by reconstructing it into a lower-dimensional latent spacethat can manage various data types. These could be particularly valuable in drug discovery as they can develop novel molecular structures (Kusner, 2017).

Augmented autoencoders such as VAE apply Bayesian advancement to help us understand possible distribution. This distribution uses input data through DL models (Kingma& Welling, 2013). The process follows the same algorithm for an autoencoder comprising encoder-decoder systems but focuses on learning probability distributions through reduced variational conclusion and recalibration. Most VAEs use neural networks and stochastic gradient descent as their basis for complex distributions that need to be understood without supervision.

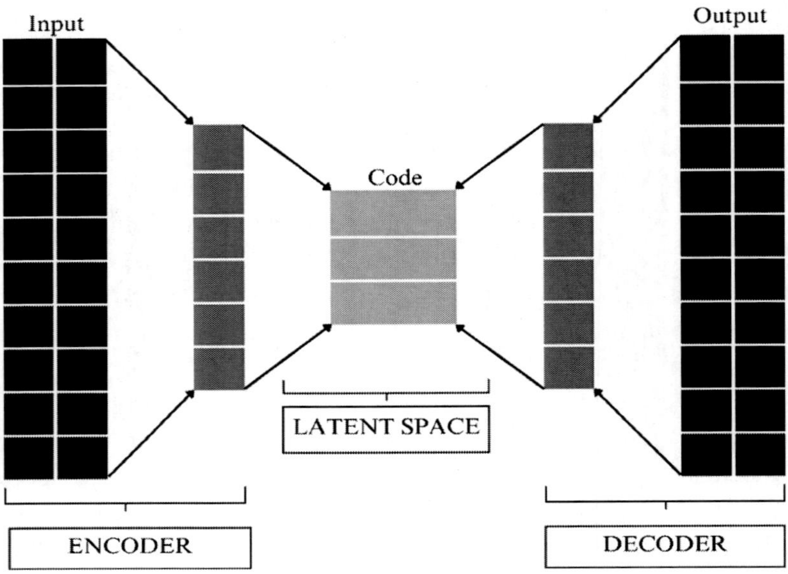

Figure 2. A schematic representation of an autoencoder.

Doerch et al. showed that various handwritten document and CIFAR image data formats can be produced using autoencoders (Doersch, 2016). Some applications in VAEs are trading, disease diagnosis (particularly concerning electrocardiogram), biomedical signal processing, generating synthetic volatility surfaces, image data augmentation, and speech recognition systems (Singh & Ogunfunmi, 2021).

Significance of Autoencoders and Variational Autoencoders in Drug Discovery

DL algorithms can customize therapeutic candidates to particular biological targets, which has drawn much interest in the pharmaceutical industry (Chen et al., 2018). DL models have been discovered to produce verbally incapacitated molecules, resulting in disputes inthe implementation of drug discovery and development (Lavecchia, 2019).

The innovative approach employs growth inhibition data and chemical fingerprints to produce medication applicants using a conditional variational autoencoder (CVAE). Even in the demonstration, when the vital chemical attributes are absent in the training dataset, our advance produces fingerprints that seize them. Furthermore, the application, by any means, canbe used for deep research purposes, enabling the use of generated fingerprints as extended queries to search databases (Joo et al., 2020).

The VAE is a more general technique that could be applied to many DL architectures,including convolutional neural networks (CNN), graph neural networks (GNN), and extendedshort-term memory models (Kutsal, Ucar, & Kati, 2024). A Denoising Autoencoder (DAE) is a significant type of variational autoencoder widely used for removing noise-corrupted information that presents random noise in the input. This process leads to the discrepancy between the original inputs and reconstructed inputs. Various types of noises, such as salt andpepper noise, Additive White Gaussian Noise (AWGN), and masking noise, can be introduced during these processes (Vincent et al., 2010).

Clustering Techniques for Compounding Analysis

Overview of Clustering Techniques

Clustering techniques are statistical data analysis approaches to unsupervised learning models that are usually used to investigate and visualize unlabeled datasets. Some specific applications of these clustering techniques in data analysis include pattern identification and extraction, block quantization (BQ), picture categorization, function approximation, and data processing (Du, 2010). In this approach, a grouping of data happens based on the properties of the datasets, where similar data points are grouped as a cluster and dissimilar data points are grouped in a different cluster, which can be achieved by using various competitive clustering methods (Du, 2010; Velunachiyar & Sivakumar, 2023). These various clustering neuralnetwork models, such as the learning vector quantization (LVQ), the self-organizing map (SOM), the ART model, and the neural gas, along with clustering algorithms such as mountain/subtractive clustering, the C-means, and fuzzy C-means (FCM) algorithms, are extensively discussed in the cited study (Du, 2010).

The widely used clustering algorithms are K-means, hierarchical clustering, and density-basedclustering (DBSCAN). These algorithms are employed to achieve the tasks mentioned above,each with its unique approach and efficiency metrics ("Clustering," 2023). The results obtained from a clustering technique can be validated by various validation methods to ensure the accuracy and reliability of the clustering outcomes. These connectivity-based and prototype- based approaches to cluster validation methods include silhouette analysis, cluster purity evaluation, external index measurements, and internal evaluation metrics analysis (Kapp- Joswig & Keller, 2022). The challenges like weak connectivity and heterogeneous density in the data associated with these analysis and validation techniques are addressed by the boundary-seeking clustering approach using local direction centrality, which has employed a density-independent metric based on K-nearest neighbors (KNNs) distribution (Peng et al., 2022). A general pictorial representation of the clustering algorithm has been shown in Figure 3.

Applications of clustering techniques are widely spread across various fields. They alsoplay a vital role in the pharmaceutical industry, helping in drug discovery and developmentprocesses, market analysis, and more. In the following section, we will discuss the importance of these clustering techniques for compound analysis.

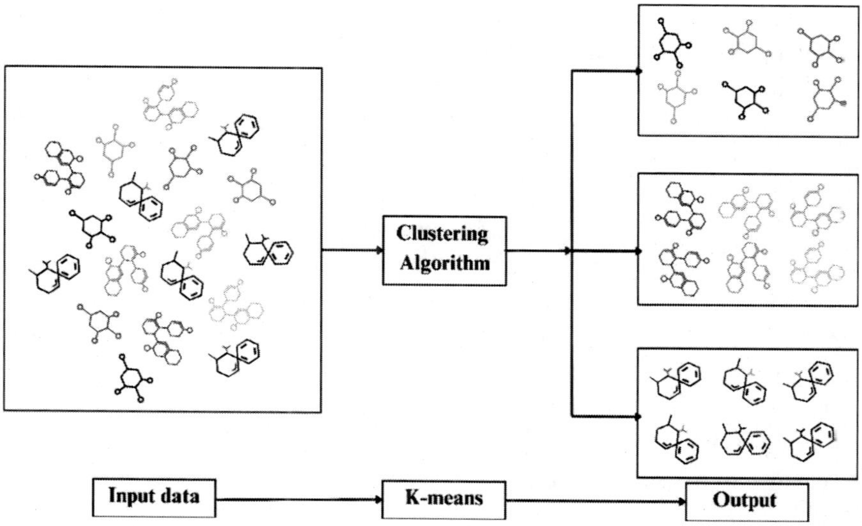

Figure 3. A general representation of the clustering algorithm. This image illustrates the grouping of molecules based on their structure similarity into distinct clusters usinga clustering algorithm.

Importance of Clustering Techniques for Compound Analysis

Clustering techniques can play an essential role in the analysis of the compounds during the CADD process. The various clustering algorithms discussed earlier in this book chaptercan be used on large datasets of chemical compounds to classify them based on their property similarities. For example, a study by Kazempour et al., highlighted the applicationof k-means and hierarchical clustering on Mol2Vec embeddings to segment molecular compounds (Kazempour et al., 2021).

Clustering algorithms help identify specific clusters of compounds and enable tasks such as selecting compounds, generating virtual libraries, QSAR analyses, and ADMET analysis. These algorithms include bisecting k-means and ward clustering, used in variousdrug discovery steps, from identifying lead compounds to toxicity studies (Malhat, Mousa, & El-Sisi, 2014). A compound's inherent structures present within a cluster could be revealed based on the proximity between molecules within the group (Mishra et al., 2020). In a larger dataset of molecules, their distinct features could be recognized by using the clustering mentioned above methods without prior knowledge of the data structure (El Abbassi et al., 2021). These methods could also help find hidden patterns by studying individual data points that might have been missed (Böcker et al., 2005; Jayasimhan & Gadge, 2012).

Clustering algorithms are also being used to identify the distinctive characteristics that differentiate between groups. These give a visual representation of the data structure and help in understanding the distribution of the similarities and differences between compounds (Engels et al., 2006). The screening efficacy of the predicted data can also be correlated with available models using clustering methods (Liu & Johnson, 2009).

Cluster analysis usually groups compounds based on their structural similarities. It is known that the properties of the compounds are generally related to their structure;therefore, when some compounds with intriguing biological functions are identified, the other compounds in that cluster are considered closely. This helps to prioritize the compounds with desired biological activity, saving time and resources. The structure of thecompound is the chemical "fingerprint," a proper clustering method reported in the cited study; namely, mclust and AutoClass-C can handle many fingerprints and massive datasets. Therefore, the clustering techniques mainly target the structural keys and molecular fingerprints, thus shunning conventional molecular descriptors (Guttiérrez Toscano & Marriott, 1999). A new avenue of clustering technique has been reported to analyze drug- drug interactions to explore patterns of drug activity (Udrescu et al., 2016). The applicationof clustering techniques is not limited to only the ones mentioned above; with the advancement of technologies, we can witness more new algorithms or approaches.

Current Challenges and Future Opportunities

Regardless of the advantages of AI in drug discovery, many obstacles and limitations need to be considered. The main concerns and critical challenges are accessibility and availability of the appropriate information (Vamathevan et al., 2019). AI-based advances usually require vast amounts of instructional data (Tsuji et al., 2021). Frequently, the amount of information that is available may be restricted, or the information provided may be of poor quality or may contain errors; this may have an impact on the outcome's dependability and correctness (Gómez-Bombarelli et al., 2018; Silvia & Carr, 2020). Another dispute is about ethical concerns, as AI-based techniques may give rise to issues with prejudice and objectivity(Basu, Engel-Wolf, & Menzer, 2020; Kleinberg, 2018). Recent AI-based processes are not substituting for traditional experimental processes; therefore, these models cannot replace human researchers' knowledge and experience (Grebner et al., 2021). Performance gains are hampered by the deeper and more experienced neural network architecture advised by currentdeep learning techniques (Liu, Zheng, & Han, 2021). Generalizing unsupervised learning models for various applications—such as visual synthesis, compression, encoding, and recognition—presents dispute because of their limited data, diversity, and complex dimensions. It provides a partial learning method for data mapping (Abukmeil et al., 2020). Overreliance on unsupervised approaches can pose substantial ethical and epistemological problems since they may create "objective" or privileged outcomes (Watson, 2023). For example, the two main obstacles facing current molecular image-based drug discovery techniques are (1) success in the absence of labels and (2) extracting chemical structures fromthe implicitly encoded pictures (Xiang et al., 2023).

With the help of AI and ML, the pharmaceutical sector is changing, overcoming data barriers, and increasing productivity. To optimize data value, translational medicine has to usethese tools to improve traditional methods and encourage cross-functional partnerships (Terranova, Venkatakrishnan, & Benincosa, 2021). Deep learning is an extension of AIbreakthroughs that substantially influence drug development by lowering costs, time, and job demands during the early phases of the process (Lavecchia, 2019). Several suggestions havebeen made to address the productivity challenges in the pharmaceutical industries, includingearly-stage target identification, a more profound understanding of biological processes, illnessmodeling, improving drug delivery techniques, and classifying patients (Gower et al., 2016). These ML algorithms outperform classical methods, can generate hypotheses, and providemechanistic assessments of the MID3 framework without making explicit assumptions(Terranova, Venkatakrishnan, & Benincosa, 2021). By combining the expertise, knowledge,and experience of human researchers with AI's predictive power (Jarrahi, 2018), we cangenerate hypotheses, assess mechanisms and ultimately accelerate the process of developmentof a new drug (Wang et al., 2019).

Conclusion

In conclusion, unsupervised learning models provide valuable approaches for analysing exploratory datasets, which helps gain new and innovative insights about complex datasets. However, it is also important to note that these approaches face numerous challenges, mainly related to the authenticity of the results. To overcome these limitations, implementing rigorous validation methods ensures the outcomes' accuracy and reliability.

References

Abukmeil, M., Ferrari, S., Genovese, A., Piuri, V., & Scotti, F. (2020, 2020/10). Unsupervised Learning From Limited Available Data by β-NMF and Dual Autoencoder 2020 *IEEE International Conference on Image Processing* (ICIP), http://dx.doi.org/10.1109/icip40778.2020.9191252.

Alizadeh Behbahani, B., Tabatabaei Yazdi, F., Shahidi, F., Mortazavi, S. A., & Mohebbi, M. (2017). Principle component analysis (PCA) for investigation of relationship between populationdynamics of microbial pathogenesis, chemical and sensory characteristics in beef slices containing Tarragon essential oil. *Microbial Pathogenesis*, *105*, 37-50. https://doi.org/10.1016/j.micpath.2017.02.013.

Artoni, F., Delorme, A., & Makeig, S. (2018). Applying dimension reduction to EEG data by Principal Component Analysis reduces the quality of its subsequent Independent Component decomposition. *NeuroImage*, *175*, 176-187. https://doi.org/10.1016/j.neuroimage.2018.03.016.

Badillo, S., Banfai, B., Birzele, F., Davydov, II, Hutchinson, L., Kam-Thong, T., Siebourg-Polster, J., Steiert, B., & Zhang, J. D. (2020). An Introduction to Machine Learning. *Clinical pharmacology and therapeutics*, *107*(4), 871-885. https://doi.org/10.1002/cpt.1796.

Baldi, P., & Lu, Z. (2012). Complex-valued autoencoders. *Neural networks : the official journal of theInternational Neural Network Society*, *33*, 136-147. https://doi.org/10.1016/j.neunet.2012.04.011.

Ban, T. A. (2006). The role of serendipity in drug discovery. *Dialogues in clinical neuroscience*, *8*(3),335-344. https://doi.org/10.31887/DCNS.2006.8.3/tban.

Basu, T., Engel-Wolf, S., & Menzer, O. (2020). The Ethics of Machine Learning in Medical Sciences:Where Do We Stand Today? *Indian journal of dermatology*, *65*(5), 358-364. https://doi.org/10.4103/ijd.IJD_419_20.

Bateman, R. M., Sharpe, M. D., Jagger, J. E., Ellis, C. G., Sole-Violan, J., Lopez-Rodriguez, M., Herrera-Ramos, E., Ruiz-Hernandez, J., Borderias, L., Horcajada, J., Gonzalez-Quevedo, N., Rajas, O., Briones, M., Rodriguez de Castro, F., Rodriguez Gallego, C., Esen, F., Orhun, G., Ergin Ozcan,P., Senturk, E., Prandi, E. (2016). 36th International Symposium on Intensive Care and Emergency Medicine: Brussels, Belgium. 15-18 March 2016. *Crit Care*, *20*(Suppl 2), 94. https://doi.org/10.1186/s13054-016-1208-6.

Becht, E., McInnes, L., Healy, J., Dutertre, C.-A., Kwok, I. W. H., Ng, L. G., Ginhoux, F., & Newell, E. W.(2018). Dimensionality reduction for visualizing single-cell data using UMAP. *Nature biotechnology*. https://doi.org/10.1038/nbt.4314.

Blanco-González, A., Cabezón, A., Seco-González, A., Conde-Torres, D., Antelo-Riveiro, P., Piñeiro, Á.,& Garcia-Fandino, R. (2023). The Role of AI in Drug Discovery: Challenges, Opportunities, and Strategies. *Pharmaceuticals*, *16*(6).

Böcker, A., Derksen, S., Schmidt, E., Teckentrup, A., & Schneider, G. (2005). A Hierarchical ClusteringApproach for Large Compound Libraries. *Journal of Chemical Information and Modeling*, *45*(4), 807-815. https://doi.org/10.1021/ci0500029.

Bourlard, H., & Kabil, S. H. (2022). Autoencoders reloaded. *Biological cybernetics*, *116*(4), 389-406.https://doi.org/10.1007/s00422-022-00937-6.

Burzykowski, T., Rousseau, A. J., Geubbelmans, M., & Valkenborg, D. (2023). Introduction to machine learning. *Am J Orthod Dentofacial Orthop*, *163*(5), 732-734. https://doi.org/10.1016/j.ajodo.2023.02.005.

Camacho, D. M., Collins, K. M., Powers, R. K., Costello, J. C., & Collins, J. J. (2018). Next-Generation Machine Learning for Biological Networks. *Cell*, *173*(7), 1581-1592. https://doi.org/10.1016/j.cell.2018.05.015.

Chen, H., Engkvist, O., Wang, Y., Olivecrona, M., & Blaschke, T. (2018). The rise of deep learning indrug discovery. *Drug Discovery Today*, *23*(6), 1241-1250. https://doi.org/10.1016/j.drudis.2018.01.039.

Chen, W., Liu, X., Zhang, S., & Chen, S. (2023). Artificial intelligence for drug discovery: Resources,methods, and applications. *Molecular Therapy - Nucleic Acids*, *31*, 691-702. https://doi.org/10.1016/j.omtn.2023.02.019.

Choi, R. Y., Coyner, A. S., Kalpathy-Cramer, J., Chiang, M. F., & Campbell, J. P. (2020). Introduction to Machine Learning, Neural Networks, and Deep Learning. *Transl Vis Sci Technol*, *9*(2), 14. https://doi.org/10.1167/tvst.9.2.14.

Clustering. (2023). In *Principles and Theories of Data Mining With RapidMiner* (pp. 139-159). IGIGlobal. https://doi.org/10.4018/978-1-6684-4730-7.ch007.

Dara, S., Dhamercherla, S., Jadav, S. S., Babu, C. M., & Ahsan, M. J. (2022). Machine Learning in Drug Discovery: A Review. *Artificial intelligence review*, *55*(3), 1947-1999. https://doi.org/10.1007/s10462-021-10058-4.

Deore, A. B., Dhumane, J. R., Wagh, R., & Sonawane, R. (2019). The Stages of Drug Discovery and Development Process. *Asian Journal of Pharmaceutical Research and Development*, *7*(6), 62-67. https://doi.org/10.22270/ajprd.v7i6.616.

Ding, J., Condon, A., & Shah, S. P. (2018). Interpretable dimensionality reduction of single cell transcriptome data with deep generative models. *Nature communications*, *9*(1), 2002-2002.https://doi.org/10.1038/s41467-018-04368-5.

Doersch, C. (2016). Tutorial on variational autoencoders. *arXiv (Cornell University)*. https://doi.org/https://doi.org/10.48550/arxiv.1606.05908.

Du, K. L. (2010). Clustering: a neural network approach. *Neural networks: the official journal of the International Neural Network Society*, *23*(1), 89-107. https://doi.org/10.1016/j.neunet.2009.08.007.

Ece, A. (2023). Computer-aided drug design. *BMC chemistry*, *17*(1), 26-26.https://doi.org/10.1186/s13065-023-00939-w.

El Abbassi, M., Overbeck, J., Braun, O., Calame, M., van der Zant, H. S. J., & Perrin, M. L. (2021). Benchmark and application of unsupervised classification approaches for univariate data. *Communications Physics*, *4*(1). https://doi.org/10.1038/s42005-021-00549-9.

Elhaik, E. (2022). Principal Component Analyses (PCA)-based findings in population genetic studiesare highly biased and must be reevaluated. *Scientific reports*, *12*(1), 14683-14683. https://doi.org/10.1038/s41598-022-14395-4.

Engels, M. F. M., Gibbs, A. C., Jaeger, E. P., Verbinnen, D., Lobanov, V. S., & Agrafiotis, D. K. (2006). ACluster-Based Strategy for Assessing the Overlap between Large Chemical Libraries and ItsApplication to a Recent Acquisition. *Journal of Chemical Information and Modeling*, *46*(6), 2651-2660. https://doi.org/10.1021/ci600219n.

Georgevici, A. I., & Terblanche, M. (2019). Neural networks and deep learning: a brief introduction. *Intensive Care Med*, *45*(5), 712-714. https://doi.org/10.1007/s00134-019-05537-w.

Giuliani, A. (2017). The application of principal component analysis to drug discovery and biomedical data. *Drug Discovery Today*, *22*(7), 1069-1076. https://doi.org/10.1016/j.drudis.2017.01.005.

Gola, D., Mahachie John, J. M., van Steen, K., & König, I. R. (2016). A roadmap to multifactordimensionality reduction methods. *Briefings in Bioinformatics*, *17*(2), 293-308. https://doi.org/10.1093/bib/bbv038.

Gómez-Bombarelli, R., Wei, J. N., Duvenaud, D., Hernández-Lobato, J. M., Sánchez-Lengeling, B., Sheberla, D., Aguilera-Iparraguirre, J., Hirzel, T. D., Adams, R. P., & Aspuru-Guzik, A. (2018). Automatic Chemical Design Using a Data-Driven Continuous Representation of Molecules. *ACS central science*, *4*(2), 268-276. https://doi.org/10.1021/acscentsci.7b00572.

Gower, N. J. D., Barry, R. J., Edmunds, M. R., Titcomb, L. C., & Denniston, A. K. (2016). Drug discovery in ophthalmology: past success, present challenges, and future opportunities. *BMC ophthalmology*, *16*, 11-11. https://doi.org/10.1186/s12886-016-0188-2.

Grebner, C., Matter, H., Kofink, D., Wenzel, J., Schmidt, F., & Hessler, G. (2021). Application of Deep Neural Network Models in Drug Discovery Programs. *ChemMedChem*, *16*(24), 3772-3786.https://doi.org/10.1002/cmdc.202100418.

Greenacre, M., Groenen, P. J. F., Hastie, T., D'Enza, A. I., Markos, A., & Tuzhilina, E. (2022). Principal component analysis. *Nature Reviews Methods Primers*, *2*(1). https://doi.org/10.1038/s43586-022-00184-w.

Gupta, R., Srivastava, D., Sahu, M., Tiwari, S., Ambasta, R. K., & Kumar, P. (2021). Artificial intelligence to deep learning: machine intelligence approach for drug discovery. *Molecular diversity*, *25*(3), 1315-1360. https://doi.org/10.1007/s11030-021-10217-3.

Guttiérrez Toscano, P., & Marriott, F. H. C. (1999). Unsupervised Classification of Chemical Compounds. *Journal of the Royal Statistical Society Series C: Applied Statistics*, *48*(2), 153-163. https://doi.org/10.1111/1467-9876.00146.

Hagele, D., Krake, T., & Weiskopf, D. (2023). Uncertainty-Aware Multidimensional Scaling. *IEEE Transactions on Visualization and Computer Graphics*, 1-10. https://doi.org/10.1109/tvcg.2022.3209420.

Hussain, A. (2019). The Next Generation Sequencing Techniques and Application in Drug Discoveryand Development. In (pp. 240-259): IGI Global.

Ijaz, M. S., Awan, M. F., Ashraf, J., Ajaz, M., Iqbal, A., & Rashid, H. (2023, 27-29 Nov. 2023). The Applications Of Artificial Intelligence In Data Science, Big Data Analytics, Cybersecurity, GIS and Nanotechnology. 2023 *2nd International Conference on Emerging Trends in Electrical, Control, and Telecommunication Engineering* (ETECTE).

Jarrahi, M. H. (2018). Artificial intelligence and the future of work: Human-AI symbiosis inorganizational decision making. *Business Horizons*, *61*(4), 577-586. https://doi.org/10.1016/j.bushor.2018.03.007.

Jayasimhan, A., & Gadge, J. (2012). Anomaly Detection using a Clustering Technique. *International Journal of Applied Information Systems*, *2*(8), 5-9. https://doi.org/10.5120/ijais12-450391.

Joo, S., Kim, M. S., Yang, J., & Park, J. (2020). Generative Model for Proposing Drug Candidates Satisfying Anticancer Properties Using a Conditional Variational Autoencoder. *ACS omega*, *5*(30), 18642-18650. https://doi.org/10.1021/acsomega.0c01149.

Kapp-Joswig, J. O., & Keller, B. G. (2022). Clustering - Basic concepts and methods. In *arXiv.org* (Vol.abs/2212.01248).

Katiyar, C., Gupta, A., Kanjilal, S., & Katiyar, S. (2012). Drug discovery from plant sources: An integrated approach. *Ayu*, *33*(1), 10-19. https://doi.org/10.4103/0974-8520.100295.

Kazempour, D., Beer, A., Oelker, M., Kröger, P., & Seidl, T. (2021). Compound Segmentation via Clustering on Mol2Vec-based Embeddings. In.

Kingma, D. P., & Welling, M. J. a. p. a. (2013). Auto-encoding variational bayes. *arXiv (CornellUniversity)*. . https://doi.org/https://doi.org/10.48550/arxiv.1312.6114.

Kleinberg, J. (2018, 2018/06/12). *Inherent Trade-Offs in Algorithmic Fairness* Abstracts of the 2018ACM International Conference on Measurement and Modeling of Computer Systems, http://dx.doi.org/10.1145/3219617.3219634.

Kusner, M. J., Paige, B., Hernández-Lobato, J. M. (2017). Grammar variational autoencoder. *Proceedings of the 34th International Conference on Machine Learning* ICML '17, https://proceedings.mlr.press/v70/kusner17a.html.

Kutsal, M., Ucar, F., & Kati, N. (2024). Computational drug discovery on human immunodeficiency virus with a customized long short-term memory variational autoencoder deep-learning architecture. *CPT: pharmacometrics & systems pharmacology*, *13*(2), 308-316. https://doi.org/10.1002/psp4.13085.

Lavecchia, A. (2019). Deep learning in drug discovery: opportunities, challenges and future prospects. *Drug Discovery Today*, *24*(10), 2017-2032. https://doi.org/10.1016/j.drudis.2019.07.006.

Liu, W., & Johnson, D. (2009). Clustering and Its Application in Multi-Target Prediction. *Currentopinion in drug discovery & development*, *12*, 98-107.

Liu, Z., Zheng, Y., & Han, X. H. (2021). Deep Unsupervised Fusion Learning for Hyperspectral ImageSuper Resolution. *Sensors (Basel, Switzerland)*, *21*(7), 2348. https://doi.org/10.3390/s21072348.

Malhat, M. G., Mousa, H. M., & El-Sisi, A. B. (2014, 2014/12). *Clustering of chemical data sets fordrug discovery* 2014 9th International Conference on Informatics and Systems, http://dx.doi.org/10.1109/infos.2014.7036702.

Mishra, B. S. P., Pandey, O., Dehuri, S., & Cho, S.-B. (2020). Unsupervised Functional Link ArtificialNeural Networks for Cluster Analysis. *IEEE Access*, *8*, 169215-169228. https://doi.org/10.1109/access.2020.3024111.

Muthuraj, R., Gopal, D., Ahmed, I., & Chandrasekaran, J. (2024). Insightful t-SNE guided exploration spotlighting Palbociclib and Ribociclib analogues as novel WEE1 kinase inhibitory candidates. *J Biomol Struct Dyn,* 1-13. https://doi.org/10.1080/07391102.2024.2305316.

Patil, S., Albogami, S., Hosmani, J., Mujoo, S., Kamil, M. A., Mansour, M. A., Abdul, H. N., Bhandi, S., & Ahmed, S. (2022). Artificial Intelligence in the Diagnosis of Oral Diseases: Applications and Pitfalls. *Diagnostics (Basel, Switzerland)*, *12*(5), 1029. https://doi.org/10.3390/diagnostics12051029.

Paul, D., Chakraborty, S., & Das, S. (2024). Robust Principal Component Analysis: A Median of MeansApproach. *IEEE Transactions on Neural Networks and Learning Systems*, 1-13. https://doi.org/10.1109/tnnls.2023.3298011.

Peng, D., Gui, Z. W., Wang, D., Ma, Y., Huang, Z., Zhou, Y., & Wu, H. (2022). Clustering by measuringlocal direction centrality for data with heterogeneous density and weak connectivity. In *Nature communications* (Vol. 13).

Pina, A. S., Hussain, A., & Roque, A. C. A. (2009). An Historical Overview of Drug Discovery. In *Ligand-Macromolecular Interactions in Drug Discovery* (pp. 3-12): Humana Press.

Polanski, J. (2022). Unsupervised Learning in Drug Design from Self-Organization to Deep Chemistry. *International journal of molecular sciences*, *23*(5), 2797.https://doi.org/10.3390/ijms23052797.

Prabha, S., Sasikumar, S., Surendra, S., Chennakeshava, P., & Reddy, Y. S. M. (2023). Drug Discovery Analysis Using Machine Learning Bioinformatics. In A. Khanna, Z. Polkowski, & O. Castillo, *Proceedings of Data Analytics and Management* Singapore.

Principal Component Analysis. (2022). In T. DelSole & M. Tippett (Eds.), *Statistical Methods for Climate Scientists* (pp. 273-297). Cambridge University Press. https://doi.org/10.1017/9781108659055.013.

Rao, M., Gupta, R., Liguori, M., Hu, M., Huang, X., Mantena, S., Mittelstadt, S., Blomme, E., & Van Vleet, T. (2019). Novel Computational Approach to Predict Off-Target Interactions for Small Molecules. *Frontiers in Big Data*, *2*, 25. https://doi.org/10.3389/fdata.2019.00025.

Santagati, R., Aspuru-Guzik, A., Babbush, R., Degroote, M., González, L., Kyoseva, E., Moll, N., Oppel,M., Parrish, R. M., Rubin, N. C., Streif, M., Tautermann, C. S., Weiss, H., Wiebe, N., & Utschig-Utschig, C. (2024). Drug design on quantum computers. *Nature Physics*, *20*(4), 549-557. https://doi.org/10.1038/s41567-024-02411-5.

Sarker, I. H. (2021). Machine Learning: Algorithms, Real-World Applications and Research Directions. *SN computer science*, *2*(3), 160. https://doi.org/10.1007/s42979-021-00592-x.

Silvia, H., & Carr, N. (2020). When Worlds Collide: Protecting Physical World Interests Against VirtualWorld Malfeasance. *Michigan Technology Law Review* (26.2), 279. https://doi.org/10.36645/mtlr.26.2.when.

Singh, A., & Ogunfunmi, T. (2021). An Overview of Variational Autoencoders for Source Separation, Finance, and Bio-Signal Applications. *Entropy (Basel, Switzerland)*, *24*(1), 55. https://doi.org/10.3390/e24010055.

Singhal, S., Chen, J., Beare, R., Ma, H., Ly, J., & Phan, T. G. (2012). Application of principal component analysis to study topography of hypoxic–ischemic brain injury. *NeuroImage*, *62*(1), 300-306.https://doi.org/10.1016/j.neuroimage.2012.04.025.

Tang, Y., Nai, W., Lei, N., Li, D., & Yang, Z. (2024, 24-25 May 2024). t-SNE Based on Halton Sequence Initialized Mountaineering Team-Based Optimization Algorithm. *2024 IEEE 14th International Conference on Electronics Information and Emergency Communication (ICEIEC).*

Taylor, G. W., Hinton, G. E., & Roweis, S. T. (2007). Modeling Human Motion Using Binary Latent Variables. In *Advances in Neural Information Processing Systems 19* (pp. 1345-1352): The MITPress.

Terranova, N., Venkatakrishnan, K., & Benincosa, L. J. (2021). Application of Machine Learning in Translational Medicine: Current Status and Future Opportunities. *The AAPS journal*, *23*(4),74-74. https://doi.org/10.1208/s12248-021-00593-x.

Tsuji, S., Hase, T., Yachie-Kinoshita, A., Nishino, T., Ghosh, S., Kikuchi, M., Shimokawa, K., Aburatani, H., Kitano, H., & Tanaka, H. (2021). Artificial intelligence-based computational framework for drug-target prioritization and inference of novel repositionable drugs for Alzheimer's disease.*Alzheimer's research & therapy*, *13*(1), 92-92. https://doi.org/10.1186/s13195-021-00826-3.

Udrescu, L., Sbârcea, L., Topîrceanu, A., Iovanovici, A., Kurunczi, L., Bogdan, P., & Udrescu, M. (2016).Clustering drug-drug interaction networks with energy model layouts: community analysis and drug repurposing. *Scientific reports*, *6*, 32745-32745. https://doi.org/10.1038/srep32745.

Vamathevan, J., Clark, D., Czodrowski, P., Dunham, I., Ferran, E., Lee, G., Li, B., Madabhushi, A., Shah,P., Spitzer, M., & Zhao, S. (2019). Applications of machine learning in drug discovery and development. *Nature reviews. Drug discovery*, *18*(6), 463-477.https://doi.org/10.1038/s41573-019-0024-5.

van der Maaten, L., & Hinton, G. E. (2008). Visualizing Data using t-SNE. *Journal of Machine Learning Research*, *9*. http://jmlr.org/papers/volume9/vandermaaten08a/vandermaaten08a.pdf.

Velunachiyar, S., & Sivakumar, K. (2023). Some Clustering Methods, Algorithms and their Applications. In *International Journal on Recent and Innovation Trends in Computing andCommunication.*

Vincent, P., Larochelle, H., Lajoie, I., Bengio, Y., Manzagol, P. A., & Bottou, L. J. J. o. m. l. r. (2010). Stacked denoising autoencoders: Learning useful representations in a deep network with a local denoising criterion. *11*(12). https://jmlr.org/papers/v11/vincent10a.html.

Wang, L., Ding, J., Pan, L., Cao, D., Jiang, H., & Ding, X. (2019). Artificial intelligence facilitates drugdesign in the big data era. *Chemometrics and Intelligent Laboratory Systems*, *194*, 103850.https://doi.org/10.1016/j.chemolab.2019.103850.

Watson, D. S. (2023). On the Philosophy of Unsupervised Learning. *Philosophy & Technology*,*36*(2). https://doi.org/10.1007/s13347-023-00635-6.

Wenderski, T. A., Stratton, C. F., Bauer, R. A., Kopp, F., & Tan, D. S. (2015). Principal component analysis as a tool for library design: a case study investigating natural products, brand-namedrugs, natural product-like libraries, and drug-like libraries. *Methods in molecular biology (Clifton, N. J.)*, *1263*, 225-242. https://doi.org/10.1007/978-1-4939-2269-7_18.

Wicks, C., Hudlicky, T., & Rinner, U. (2021). Chapter Two - Morphine alkaloids: History, biology, and synthesis. In H. J. Knölker (Ed.), *The Alkaloids: Chemistry and Biology*

(Vol. 86, pp. 145-342). Academic Press. https://doi.org/https://doi.org/10.1016/bs.alkal.2021.04.001.

Wold, S., Esbensen, K., & Geladi, P. (1987). Principal component analysis. *Chemometrics and Intelligent Laboratory Systems*, *2*(1-3), 37-52. https://doi.org/10.1016/0169-7439(87)80084-9.

Xiang, H., Jin, S., Liu, X., Zeng, X., & Zeng, L. (2023). Chemical structure-aware molecular image representation learning. *Briefings in Bioinformatics*, *24*(6). https://doi.org/10.1093/bib/bbad404.

Yang, Z., Shi, Y., Li, P., Pan, K., Li, G., Li, X., Yao, S., & Zhang, D. (2022). Application of Principal Component Analysis (PCA) to the Evaluation and Screening of Multiactivity Fungi. *Journal of Ocean University of China : JOUC*, *21*(3), 763-772. https://doi.org/10.1007/s11802-022-5096-x.

Yarali, A. (2023). Artificial Intelligence. In *From 5G to 6G* (pp. 149-157).https://doi.org/10.1002/9781119883111.ch9.

Yoo, C., & Shahlaei, M. (2018). The applications of PCA in QSAR studies: A case study on CCR5 antagonists. *Chemical Biology & Drug Design*, *91*(1), 137-152. https://doi.org/https://doi.org/10.1111/cbdd.13064.

Yuan, H., Tang, Y., Hao, H., Zhao, Y., Zhang, Y., & Chen, Y. (2022). Intelligent Detection Method of Gearbox Based on Adaptive Hierarchical Clustering and Subset. *Computational intelligence and neuroscience*, *2022*, 6464516. https://doi.org/10.1155/2022/6464516.

Zhang, C., Fu, H., Hu, Q., Zhu, P., & Cao, X. (2016). Flexible Multi-View Dimensionality Co-Reduction. *IEEE Transactions on Image Processing*, *PP*, 1-1. https://doi.org/10.1109/TIP.2016.2627806.

Zhang, S., Xiao, K., Carranza, E. J. M., Yang, F., & Zhao, Z. (2019). Integration of auto-encoder network with density-based spatial clustering for geochemical anomaly detection for mineral exploration. *Computers & Geosciences*, *130*, 43-56. https://doi.org/10.1016/j.cageo.2019.05.011.

Zhou, H., Wang, F., & Tao, P. (2018). t-Distributed Stochastic Neighbor Embedding Method with the Least Information Loss for Macromolecular Simulations. *J Chem Theory Comput*, *14*(11), 5499-5510. https://doi.org/10.1021/acs.jctc.8b00652.

Chapter 6

Machine Learning in Pharmacogenomics: From Target Identification to Clinical Trials

A. Ashwini[1,*]
and P. M. Ansho[2]
[1]Department of Electronics and Communication Engineering,
Vel Tech Rangarajan Dr. Sagunthala R&D Institute of Science and Technology,
Avadi, Chennai, Tamil Nadu, India
[2]Department of EEE, Cape Institute of Technology. Kanyakumari, Tamil Nadu, India

Abstract

The development of ML has played a very profound role in the advancement of pharmacogenomics, enhancing the precision and efficiency of drug identification and development. This chapter provides insights into how pharmacogenomics makes a change, starting with target identification. ML predicts targets far beyond any other technology, owing to its capability to detect genetic mutations and their connections to diseases. These computational techniques allow the identification of biomarkers and genetic risks to tailor the therapy and adapt it according to the traits of patients. The adaptation of the ML in this first stage provides not only an improvement in the speed to identify but also in the overall cost since it eliminates more costly processes such as experimentation. The chapter discusses the identification of targets in disease mechanisms, and how ML uses the fine-tuning of product leads in drug discovery. Using sophisticated statistical analysis, ML is capable of predicting the ability of a compounded drug to interact with the desired biological marker. It makes it possible to screen many compounds within a shorttime from picking out the most potent therapeutic, leading to

* Corresponding Author's Email: a.aswiniur@gmail.com

In: Harnessing AI for Drug Discovery
Editor: Kolla Bhanu Prakash
ISBN: 979-8-89530-540-9

adverse effects. In this context, the adoption of ML primarily filters through potential prospects by identifying only those possibilities which are most likely to subsequently move on to various experiments. Moreover, viability and optimizing algorithms from ML can also fine-tune drug dosages and formulations for better results. The final sub-sections of the chapter provide information on specific aspects of clinical trials involving the use of ML. In the case of trials, these ML models could be used to shift through patient profiles for candidates most appropriate for trials and hence would increase the efficiency of trials. They can track and forecast patient responses providing a capability to modify the trial. When a sufficient amount of clinical data is collected, ML looks for patterns and relational associations that are not obvious to conventional approaches. It provides a more comprehensive and conclusive trial. Utilizing ML in clinical trials also enhances the efficiency and effectiveness of clinical trials and, hence boosts the chances of success in producing new and innovative treatments. Lastly, this chapter gives the readers an insight into multiple areas of pharmacogenomics that have been enhanced by machine learning, leading to prospective medicine advancements.

Keywords: clinical testing, drug discovery, machine learning, medical care, pharmacogenomics, predictive modelling, target identification

Introduction to Pharmacogenomics and Machine Learning

Pharmacogenomics is an integration of pharmacology and genomics that focuses on the role of personal genetic makeup on drug reactions [1]. It is a new field of study that creates proper medicines and specifies appropriate doses for every person depending on their genes, thus reducing the side effects of the medication and increasing the drug's effectiveness.

For example, using ML, genomic sequences to ascertain biomarkers can be derived, that would determine a patient's response to certain drugs and therefore recommend customized treatment. Third, it supports drug design and clinical trial optimization by creating models of how drugs interact with targets, estimating side effects, and other factors. The use of ML technology in pharmacogenomics produces not only faster results but also improved accuracy in pharmacogenomics. It provides better and safer healthcare solutions. Figure 1 shows the integration of pharmacogenomics, with artificial intelligence.

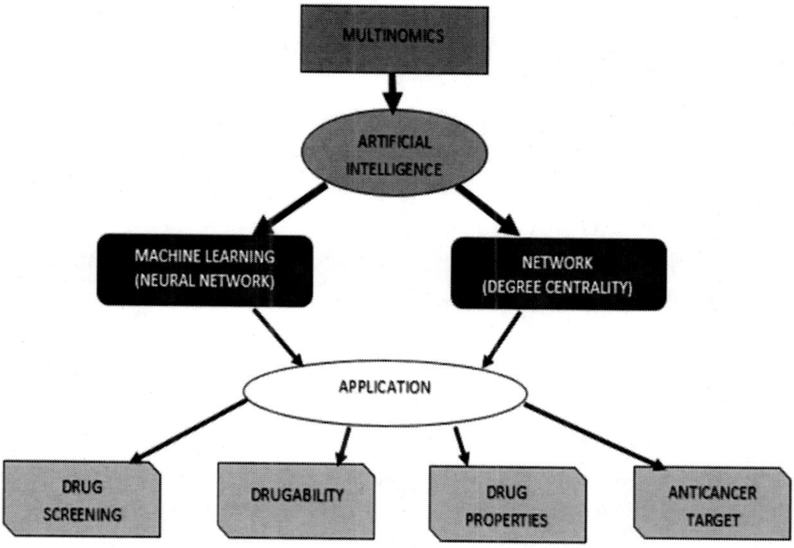

Figure 1. Artificial Intelligence Integrated with Pharmacogenomics.

Types of Data Sources in Pharmacogenomics

As in pharmacogenomics, the discovery of such variations categorizes a vast range of data to assess the role of genetic differences in the outcome of medicines. Whole genome sequencing and Whole Exome Sequencing are two techniques that are the most complex and informative about an individual's pharmacogenomic profile since they can register both common and rare genetic variations that can participate in Drug Metabolism and Response [2]. Quantitative gene panels differ from qualitative gene panels because only a targeted gene sequence is sequenced rather than genome. Sequencing of high-risk genes commonly involved in metabolism of drugs makes it suitable for clinical use where genetic analysis is time-dependent.

Transcriptomics includes RNA sequencing (RNA-Seq) and microarray analysis which provides information on gene expression. It provides knowledge on how a specific treatment interferes with gene regulation. The Methylation data and the data regarding histone deacetylation give another layer of regulation and a comparative idea of how epigenomic changes may affect the concerned genes. This may modulate the response to drugs. Gene expression profiling based on Genotyping microarray assays single nucleotide polymorphisms and copy number variations and data of proteomics through mass spectrometry. The protein microarray exhibit protein level change in the response to the drug. There is another type of omics data including metabolomic generated through nuclear magnetic resonance spectroscopy. Metabolomics based on mass spectrometry gives a picture of the metabolic pathways and the changes in the metabolite profiles due to the genetic background and drugs. These imaging and genotyping results, in combination with treatment records and such physiologic characteristics as the EHRs, establish a connection between genetics and advantageous and adverse effects, where it is possible to identify if there is a link between genetic alteration and how a particular drug works. Several databases from TCGA, PharmGKB, and 1000 Genomes Project are helpful. It contains valuable information where species variations and their impacts are categorized according to genera, sub-species, and populations [3]. All these various types of information sources together offer a more comprehensive picture of genetic determinants of metabolism and response to drugs. It makes the work on the concept of individual pharmacology possible.

Machine Learning Techniques

There is an increasing importance of AI approaches in drug design because of the capability of AI to analyze large sets of data. It helps to understand the interaction and predict the action of drugs with receptors and optimize the drug design process [4]. Figure 2 shows the general structure of a neural network structure with machine learning architecture.

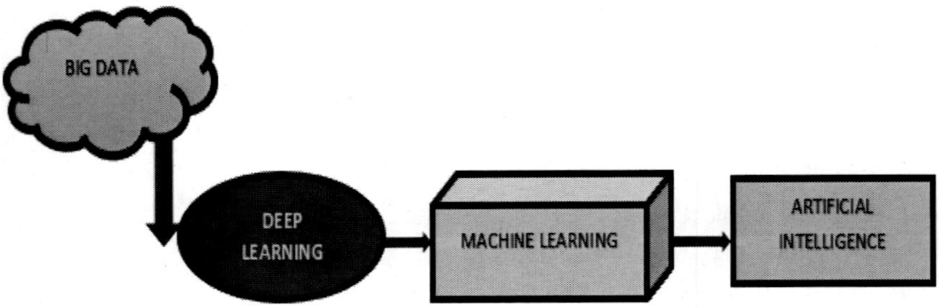

Figure 2. Structure of Machine Learning with Artificial Intelligence.

Supervised Learning

Classification Algorithms

Random Forests is applied to predict a given compound sample to be active or inactive against a selected target using molecular descriptors. A suitable example for SVM is used in predicting drug-target interactions where the best-dividing hyper-plane is found by establishing a boundary between active and inactive compounds.

Regression Algorithms

Linear Regression estimates quantitative structure-activity relationship (QSAR) which is the prediction of the biological effect of a compound based on its chemical composition. Neural Network models have the ability to map intricate, nonlinear associations between molecular attributes and biological activities for the prospects of efficacy in estimation of drugs [5].

Unsupervised Learning

Clustering Algorithms

Compounds are clustered into the same group using K-means Clustering according to their physicochemical properties. It helps in assisting the development of new drugs. Hierarchical Clustering can observe the similarity and divergence in compound space, which can be especially helpful for designing and organizing chemical libraries.

Dimensionality Reduction

PCA is used in compressing large sets of data by making them smaller in dimension while still retaining the variance of the large data set. It is easier to interpret data that contain dimensions [6]. T-Distributed Stochastic Neighbor Embedding (t-SNE), compares compound relationships primarily for analysis.

Deep Learning

Deep Learning Models, especially Convolutional Neural Networks (CNNs) are well suited for image data analysis of 3D structures. It is applied in virtual screening to determine the probability of the screening compounds interacting with their target. The primary technique is Recurrent Neural Networks (RNNs) which works well with the sequence data like protein folding and interactions for the given amino acid sequences. RNNs as mentioned earlier are special types of neural networks that are commonly used for predicting time-series data. The drug-response uses LSTM network for determining the effects of the drug over a period. Autoencoders are utilized for both the unlabeled or unlabeled feature learning and also for feature extraction where the number of features that are relevant, crucial or interesting is less than the total number of features amidst the data set. Depending on the problem, it can find latent forms in the chemical database and help identify new drug targets.

Generative Models

Generative Adversarial Networks (GANs)

GAN comprises two deep learning neural networks, namely the generator network and the discriminator network, which is used to create new realistic drug-like molecules [7]. It is used in the rational design of new drugs, drugs with characteristics satisfying specific criteria.

Variational Autoencoders (VAEs)

Encode the input data on a desired latent space while reconstructing and generating new molecular forms. It can be used for browsing the chemical space and producing new molecules to examine the initial biological activity of the candidates.

Reinforcement Learning

Deep Reinforcement Agents can be used to make a sequence of decisions over time, by taking the decisions in a trial-and-error fashion to maximize the obtained reward. Solvent selection, deciding upon the scale for a reaction, and designing new target molecules are some areas where it can be applied [8].

Ensemble Methods

Depending on the task accomplished, Boosting Algorithms such as Gradient Boosting Machines (GBM) involve using several non-strong learners to build a strong learner. It employs the prediction of drug-target interactions and possible side effects. Bootstrap Aggregating (Bagging) and Random forests are some examples of bagging technique which enhances model reliability. It is less prone to overfitting [9]. It is used in different steps in the drug discovery process to make sure that the discoveries made are accurate.

Transfer Learning

Transfer Learning, also referred to as Pre-trained Models, takes knowledge from the large dataset and applies it to a limited dataset. The algorithm is helpful where sample data is scarce, such as the prediction of response to drugs in rare illnesses. By using these related machine learning techniques, the researchers will be able to avoid side round in the drug discovery process and eventually lower the costs while improving the success rates for drug candidate identification.

Target Identification in Pharmacogenomics

Genome-Wide Association Studies (GWAS)

GWAS plays a vital role in pharmacogenomics because they offer an effective method of discovering genetic loci relevant to drug response phenotypes. Doctors perform GWAS, where they contrast the DNA of those who respond well to a drug and those who have the terrible side-effects of the drug. GWAS is a process that investigates a large population of individuals by focusing on millions of genetic markers called SNPs to identify specific sources of genetics that link with variations in drug metabolism [10]. These identified variants of genes can be used as drug metabolizing enzymes, drug transporters or the drug target itself. GWAS offers pivotal evidence on the genetic contributions towards variations in the response to medicines across individuals and genetic targets for validation and implementation. Figure 3 shows the pharmacogenomics in drug discovery and development.

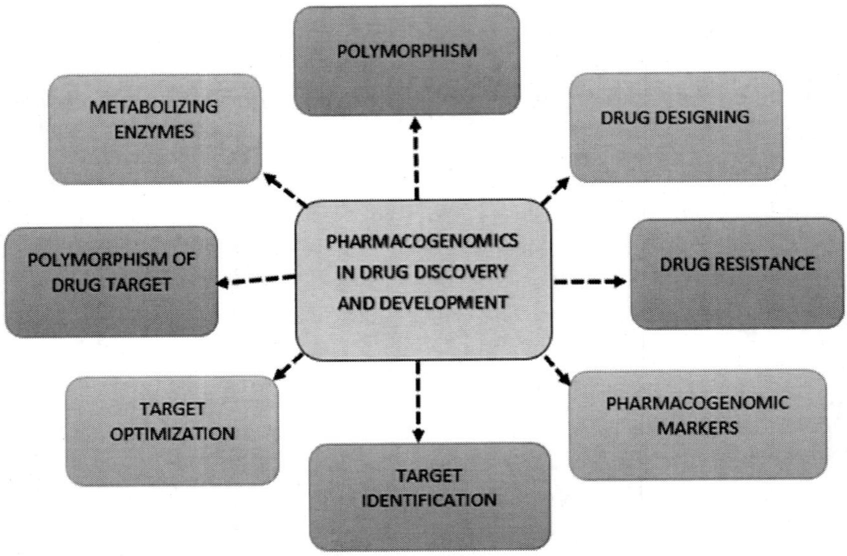

Figure 3. Pharmacogenomics in drug discovery and development.

Candidate Gene Approach

Candidate gene association specifically targets genes identified prior and which are known to be involved in the metabolic pathways of the drugs or interactions with drug targets. It is a more refined approach than the estimative method which looks at the whole genome. It compares the entire DNA makeup to the drugs, assuming that specific genes must have an effect since their function has some pharmacological relevance [11]. In general, the candidate genes choses pharmacokinetic or pharmacodynamic properties – with the hope of determining the genetic influences on drug-responsiveness, toxicity, or pharmacokinetics within a gene or locus [10]. Originally developed for studying known drug-target pathways, the candidate gene approach can give specific advice about 'Personalized Medicine' where genetic variations are connected to clinical phenotypes.

Systems Biology Approaches

Biology strategies combine genomic information with signaling and interaction networks that underlie target identification and validation. These employ algorithms and mathematical representations of gene regulatory networks, protein-protein/ligand interactions, and metabolite interconnectivity in drug target pathways. Thus, systems biology focuses on the nodes and connections influenced by genetic variations. It offers insights into how such variations affect pharmacology, toxicity, and inter-individual variability. This systems biology approach looks at broader patterns of organization of the biological systems and as a result offers a strategy for discovering new targets for drugs [11, 12]. Another advantage that systems biology has over other approaches comes to pharmacogenomic traits revolves around the fact that this approach quickly identifies interactions and dependencies.

Functional Genomics

Pharmacogenomics aims to define the effects of genotypes on the expression of proteins, the effects of drugs on cells, and other related activities. To assess how particular genetic changes affect the disease pathways and treatment response, researchers employ methods of gene expression profile analysis, RNA interference (RNAi), and CRISPR/Cas9-based screening. Leveraging genome-wide or gene-specific manipulations in cellular contexts, researchers can define drug sensitivity or resistance profiles of specific genes and pathways [13]. Functional genomics offers data and tests the functional consequences of genetic polymorphisms. The choice of targets for pharmacist intervention, functional genomics provides general guidance for new targets. A therapeutic paradigm differs from the one that resulted from the pursuance of the genome.

Computational Approaches

Target prediction strategies in target identification rely on comprehending the interplay between false positives and false negatives in drug-target prediction models, a machine learning approach, and data integration methods for selecting promising drug targets. These approaches use enormous genomic, transcriptomic, and proteomic data to find regularities and associations that connect genetic alterations with drug-reactant phenotypes. Random forests, support vector machines, and deep learning structures are used in such models for predicting drug success rates, toxicity, and drug metabolism based on gene features. These methods enhance the discovery of new targets for the drugs because these methods can process and analyze biological information quickly and in large amounts. It aids in the development of personalized medicine and better drug development lines.

Pharmacogenomic Databases and Resources

There are also public databases and resources specialized in pharmacogenomics that offer quality data regarding genetic variants, drug effects, and clinical results, including PharmGKB and disease-specific databases, for example, TCGA. These resources provide detailed databases that include both genome and phenotype data gathered from pharmacogenomic investigations and clinical studies. Scientists employ these databases to analyze trends of genetic variations concerning the reaction to medications among multiple populations and to discover new drug targets, as well as to confirm clinical trials [14]. In pharmaceutical contexts, pharmacogenomic databases are valuable resources that enable knowledge sharing and enable the growing field to progress by assisting the accommodation of discoveries and information drawn from that space.

Predictive Modeling for Biomarker Discovery

As a forward-looking medical approach, predictive modeling is indispensable in biomarker discovery because it incorporates statistical and machine learning methods to identify and verify biomarkers for the diagnosis, prognosis, and treatment response of diseases.

Data Collection and Pre-Processing

Accurate predictions for diseases and therapeutic responses require acquiring multifaceted datasets containing clinical, genomic, proteomic, or metabolomic information from patient populations or experimental sets. Data pre-processing including normalization, feature selection, deals with missing data so that the resultant data is credible [15]. A key phase in this process is the preparation phase which can significantly influence the level of accurate prognosis and criteria for biomarker identification.

Feature Selection and Dimensionality Reduction

Machine learning in biomarker discovery uses a predictive modeling approach, which consists of choosing features (variables) that are highly associated with the biomarker in large data sets. Other techniques such as principal component analysis (PCA), recursive feature elimination (RFE), or the LASSO (Least Absolute Shrinkage and Selection Operator) are applied to reduce the features set but keep essential features. These methods benefit model interpretation and decrease computational cost because the biomarker candidates' interactions are combinatory.

Model Selection and Training

Unlike other machine learning tasks, in biomarker discovery different types of algorithms for predictive modeling are utilized depending on the type of the biomarker discovery problem [16]. Popular procedures of supervised ML include support vector machine, random forest, logistic regression, and neural networks to build models, or classifiers, that can distinguish between phenotypes or prognosis agents based on biomarkers and clinical characteristics. Various techniques may be used for model selection to arrive at the best algorithm of a biomarker as would be preferred by cross-validation.

Biomarker Validation and Evaluation

In the process of biomarker-based predictive models, the model is validated to assess the accurate accuracy of the projections made. The methods used in validation are the internal methods which include splitting data into training and testing data; external methods which apply the model in a different dataset; and the registration of data and testing it iteratively known as the cross-validation type [17]. The following quantitative measures are helpful in evaluating the performance of a model and the effectiveness of biomarker predictions: From diagnostic tests, performance indicators can include sensitivity, specificity, accuracy, area under the curve (AUC), and precision-recall curves.

Interpretation and Biomarker Discovery

The advent of predictive modeling brings about more than the simple identification of biomarkers; predictive modeling provides extra value to the existing biomarkers' biological relevance and the stability of the forecast made. The assessment and analysis of features as well as the use of model interpretations applied within these studies include the combination of feature contributions. It examines the integration of specific biomarkers into predictive models and biological processes associated with a disease or treatment response [18]. This sort of analysis assists with finding new biomarkers and something about the disease process in cancer.

Clinical Translation and Application

This model is based on the optimal biomarker prediction models, which are used to create further conversion and biomarker utilization in the healthcare system. It means that through appropriate biomarkers, management of therapy, classification and results of treatments can be done. The discovery of the biomarkers is accompanied with the predictive model, the researchers move nearer to the goals of personalized medicine and ensure that treatment outcomes and prognosis improves new tools and treatment plans [19]. Computational tools, and predictive modeling enhances biomarker discovery through parsing of large datasets, selection of appropriate biomarkers, testing of the biomarkers' predictive potential and deploying the outcome in the framework of individualized patient management.

Optimization of Drug Efficacy and Safety

Optimization of drug efficacy and safety profiles are complex tasks while selecting and designing the efficacy and safety of medicines. It includes the process of discovery, development and usage of drugs till clinical applications. Here are vital aspects and strategies involved in optimizing drug efficacy and safety:

Target Identification and Validation

The strategies start with defining and then proving targets for treatment that have a critical implication in disease progression. It has less impact on other targets. This entails integrating pharmacogenomics, systems biology and computational modeling to find targets. It is important in the disease process [20].

Rational Drug Design and Screening

Molecular modeling can be defined as a successful drug design strategy as it uses structural biology, computational chemistry, and bioinformatics to create molecules that bind to specific therapeutic targets. Molecular modeling such as computer-aided drug design (CADD), molecular docking, and QSAR modeling provide an optimal binding site with high affinity and selectivity towards drug receptors, reducing off-target interactions leading to unwanted side effects [21]. High-throughput screening (HTS) is a process of screening a large number of compounds to identify the compound that has the desired pharmaceutical activity. Both automated assays and robotic systems help to augment the overall drug discovery process.

Pharmacokinetic and Pharmacodynamic Profiling

Painstaking information on a drug's PK profile assists in selecting the appropriate dosage regimen and therefore equalizing the measurement of drug exposure among patients, thus improving the treatment efficiency. Biomarkers, and surrogate markers help measure the responsiveness to therapy and estimate subsequent clinical improvements, when selecting the dose of the drug and controlling its effectiveness.

Personalized Medicine Approaches

Pharmacogenomics reveals specific gene factors impelling patients' reactions to medications to assist in creating individualized treatment plans. Through genetic testing, it is possible to anticipate the efficacy of a drug, the toxicity of the said drug to an individual, as well as the appropriate drug dose. It enhances therapeutic benefits and reduces the side effects on the same.

Preclinical and Clinical Development

Contrary to clinical trials, preclinical tests efficiently determine the safety and performance of new drugs in addition to the pharmacological actions in animals and tissue cultures. Piggybacking on strong preclinical data is a determining factor in taking a specific drug candidate to the subsequent phases of clinical trials based on safety and efficacy standards. Drug trials examine effectiveness and adverse effects in human groups, efficacy markers, and extended monitoring of adverse effects [22]. The arguments include flexible trial and biomarker strategies that increase trial efficiency and patient categorization, leading to faster progression of potential drug candidates to clinical use.

Post-Marketing Surveillance and Pharmacovigilance

Some of the advantages of post-marketing surveillance include tracking the safety and efficacy of the drugs when used in the community and can detect adverse events that were not known before the medicines launches into the market. Pharma-co-vigilance systems regularly monitor and control risks connected to marketed drugs, as well as operate to maintain patient security consistently with the assistance of administration control.

Data Integration and Computational Modelling

It achieves diverse sources, involving genomic data, electronic health records (EHRs), patient-level real-world evidence (RWE), and aggregative data that provide better insights into drug response across populations [23]. Molecular and bioinformatics methods, including machine learning system biology, implement large data sets to estimate drug interactions, provide an optimal treatment schedule for patients with disease, and select individual genetic subtypes that could receive maximum benefit from a course of therapy. The drug works are ensured as intended with minimal, acceptable or preferably no side-effects. The ultimate goal of rational drug design can only be achieved by an encompassing holistic approach from target identification to drug development to post-marketing surveillance [24]. Computer simulations and data analyses have shown the potential of pharmacogenomics, pharmacogenetics, and correlation in improving the health of people through individualized treatments by using therapeutic goals, risk factors, and bases in modern molecular biology.

Machine Learning in Drug Formulation and Dosage Assessments

Machine learning (ML) in medicine has offered an essential innovation in the structure of drug formulation and dosage estimations, using computational algorithms in the process of drug formulation, and individual patient treatment plans. These algorithms employ different mathematical models such as regression, neural networks, and genetic algorithms to establish sets of active pharmaceutical ingredients and excipients of the drug, which improve stability, solubility, and bioavailability. Immense potential for ML in designing enhanced drug delivery systems like nanoparticles focuses on controlling factors like particle size, surface charge, and drug release. This is to ensure that drug release and entire pharmacological action are selective and non-toxic to another organ.

In dosing assessments, ML models use pharmacokinetic and pharmacodynamic profiles to adjust dose levels based on the patient's traits. Population PK and physiologically based PK on mathematical algorithms model drug action in the human system depending on factors such as age, weight, genetic make-up, and disease state to predict drug concentrations over time. These predictive models allow the clinicians to select an appropriate dosing schedule. This can reduce the possibilities of adverse side reactions [25]. Furthermore, ML in contextualized medication addresses the inter-individual variations in drug response by employing genomic data and biomarkers to anticipate each patient's response to medications to tailor the appropriate dosages to optimize treatment outcomes and minimize response variability. In conclusion, the use of ML in formulation and dosage assessments not only enhances the drug development process but also arrives at better strategies that could help the patient's condition and mold the healthcare system.

Real-Time Monitoring and Adaptive Protocols

A timely response is critical in drug discovery. The change in clinical studies makes them almost real-time, flexible, and much less prone to failure.

Real-Time Monitoring

Real-time monitoring in drug discovery entails studying data at one or more time intervals in the various stages of drug development. Combined with electronic data capture (EDC), this approach quantifies patient responses, adverse events, and efficacy outcomes in real-time in clinical trials. The use of technologies like EHRs, wearable devices and remote monitoring devices, investigators can capture data in real-time and identify emerging trends, or patterns that can cause variations in trials. Real-time monitoring also has benefits that increase quality control over the particular experiment – it guarantees compliance with a specific set of protocols and standards and allows making the required corrections immediately, if necessary [26]. Furthermore, participants' safety issues and any other unforeseen reactions can be addressed and prevented promptly by developing appropriate protections for participants and the trial as a whole.

Adaptive Protocols

Adaptation in drug discovery refers to changes made during the conduct of a study, dosing schedule, patient selection in light of information gathered during the study process. Adaptive design is a concept used in clinical trial design whereby the plans of the trial can be adjusted using randomization process, dose range, or inclusion criteria. For instance, adaptive randomization could increase the number of patients to treat arms with better efficacy signals regardless of the initial randomization. It also improves the utilization of resources and expedites the search for effective treatments. Flexible Dose-Escalation Schemes strive to attain maximum therapeutic effect on the generally healthy study patients while at the same time avoiding harm by reviewing dose info accrued within a trial. Apart from increasing the trial efficiency, these adaptive strategies also increase the likelihood of finding the best approach to treat specific disease types for particular patients or at least more suitable attempts than a non-adaptive approach. Table 1 shows the different adaptive protocols used in pharmacogenomics.

Table 1. Comparison of different adaptive protocols

Protocol	Description	Merits	Demerits
Companion Diagnostics	Identification of patients through genetic markers	Improves therapeutic outcomes with the regulatory support system	The cost is high. Moreover, approval is required for both drug and diagnostics
Biomarkers	Identification of genetic markers with precise prediction and response to specific drugs	Drug response rates with the reduction in adverse reactions	Extensive biomarker discovery and privacy concerns
Genotype dosing	Adjustment of drug dosage concerning patient profile involved in drug metabolism	Reduces adverse effects, personalized treatment	The initial cost gets increased.
Precision medicine trials	Genetic information to match patients' treatment	Potential likelihood for treatment	High complexity in regulatory with large genetic data requirements.

Benefits and Applications

Several prospective advantages and use cases focus on incorporating real-time tracking and reactive approaches in drug discovery. They also increase trial efficiency because one can implement modifications based on emerging data patterns promptly, save study time and increase the speed to implement the research results into evidence-based practice. Such adaptations allow the researchers to focus on the nature of the disease and patients' differences. Flexibility paves the way to make the correct adjustments and enhance the outcomes. Additionally, the readiness of data collection in real-time helps in the management of risks and decision-making. The FDA and EMA rely on adaptive trial design, as they are aware of the benefits they present to both scientific objectivity and the potential for success, and offer regulations and structures to ensure the ethical use of clinical trials within the drug development process. Real-time and adaptable strategies mean significant changes in the drug discovery process since the primary goal is to learn without stopping, make decisions rapidly, and individualize therapeutic strategies. These approaches will help improve the original methods used in clinical trials to optimize the benefit of furthering precision medicine and the overall quality of patient treatment.

Challenges and Future Directions

Even as drug development continues to evolve in results, risks, and probabilities through technologies such as real-time monitoring and adaptive protocols, drug discovery encounters more difficult issues. Future developments strive to get around these complexities and fashion-improved results. Figure 4 shows the challenges and future directions in drug development based on machine learning.

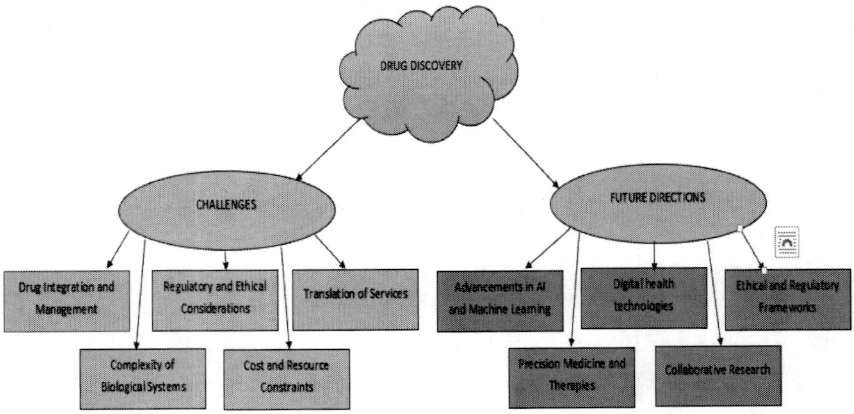

Figure 4. Challenges and Future directions in drug discovery.

Challenges

- *Data Integration and Management:* The seamless combination of genomic, clinical, and RWE remains a hurdle is primarily attributed to data heterogeneity, data structures, and stakeholders' privacy. In contrast, structured data are often accessible to handle and analyze, most of the information relevant to drug discovery is unstructured, thus requiring harmonization and standardization across various platforms to facilitate integration and analysis.

- *Complexity of Biological Systems:* Biological systems are always complicated, gene – proteins and pathways are the nested loops of biological systems. Understanding these dynamics necessitates the usage of sophisticated computational methods as well as experiments to uncover finer details of disease patterns.
- *Regulatory and Ethical Considerations:* Industry-specific governing bodies establishing rules and guidelines for drug development can be difficult and may have implementing issues like adaptive trial designs. A major concern is the need to maintain a balance between innovations and patient safety, privacy, and ethics.
- *Cost and Resource Constraints:* The search for new drugs and their development involves a great time and a significant amount of money. For smaller research organizations and other start-ups, the issues are related to funding, development of better structural support and skilled human capital.
- *Translation from Bench to Bedside:* The translation of these research studies from the preclinical stage to clinical service remains a core issue. Numerous promising drug candidates do not successfully pass through clinical trials, partly due to lack of efficacy or safety concerns, or due to patients' response. Therefore, there is a profound need for further improvement of preclinical models and biomarkers.

Future Directions

- *Advancements in Artificial Intelligence and Machine Learning:* AI & ML go hand in hand which presents continued innovation in data analysis, prediction of drug responses, and selection of efficient therapeutic strategies. AI-driven platforms can be used to analyze multiple types of data and implement personalize medicine strategies.
- *Precision Medicine and Personalized Therapies:* Personalizing a treatment plan is still the way forward including targeting genetic data, biomarkers, and disease specifics. Innovations in genomic sequencing and biomarkers are initiating new treatment options aimed at achieving therapeutic efficiency and avoiding toxicity.
- *Digital Health Technologies:* People use wearable devices, mobile health applications, and remote monitoring systems to collect data

from the patients in real-time clinical trial processes and allow continuous patient tracking.

- *Collaborative Research Initiatives:* Data sharing, standardization and regulatory issues can be addressed by affiliating the formation of a trilateral collaboration among the academia, the industry, and the regulating agencies. Consortiums and public-private partnerships entail new, and moral ideas that help with knowledge sharing.
- *Ethical and Regulatory Frameworks:* It is crucial to revise the existing regulatory standards and ethical codes to reflect the new trial designs, protect patients' rights and sooner approve the drugs and medical devices with a high potential. Such an approach to regulation, including the use of adaptive regulatory pathways and properly constructed guidelines creates new scientific developments.

Conclusion

In conclusion, drug discovery as a field is in the midst of a revolutionary shift, which is possible due to the incorporation of new technologies, reliance on data, and a movement toward personalized medicine. Because AI plays a vital role in optimizing care to match patient's needs where medication and treatments are more specific to the patient's needs. Some problems that must be overcome which cannot give clear solutions.

AI can find pairings between patients and their corresponding medications that meet the highest probabilities of success in the tumor microenvironment. Clinical decision support systems enhance health care because they make recommendations on effective treatment for their patients based on information fed to them by doctors. This method improves the therapeutic outcomes, reduces the manifestations of side effects in the patient's body, and increases the satisfaction levels of the patient. AI-driven PP is arguably one of the most promising approaches to reboot healthcare systems. Maintaining patient confidence and rapport can be built only if their data is protected. Increased numbers of collaborative research programs, on patents and clinical trials, help to speed up the move from laboratory to clinic. Altogether, as these developments progress, they indicate that drug development is not only proceeding at a faster pace but also that today's standards of treatment are becoming inadequate for the future, which should deliver even better results and help more people live better lives.

References

[1] Kalinin, A. A., Higgins, G. A., Reamaroon, N., Soroushmehr, S., Allyn-Feuer, A., Dinov, I. D. & Athey, B. D. (2018). Deep learning in pharmacogenomics: from gene regulation to patient stratification. *Pharmacogenomics*, *19*(7), 629-650.

[2] Kalamara, A., Tobalina, L., & Saez-Rodriguez, J. (2018). How to find the right drug for each patient? Advances and challenges in pharmacogenomics. *Current opinion in systems biology*, *10*, 53-62.

[3] Ryan, D. K., Maclean, R. H., Balston, A., Scourfield, A., Shah, A. D., & Ross, J. (2024). Artificial intelligence and machine learning for clinical pharmacology. *British Journal of Clinical Pharmacology*, *90*(3), 629-639.

[4] Burczynski, M. E., Oestreicher, J. L., Cahilly, M. J., Mounts, D. P., Whitley, M. Z., Speicher, L. A., & Trepicchio, W. L. (2005). Clinical pharmacogenomics and transcriptional profiling in early phase oncology clinical trials. *Current Molecular Medicine*, *5*(1), 83-102.

[5] Burczynski, M. E., Oestreicher, J. L., Cahilly, M. J., Mounts, D. P., Whitley, M. Z., Speicher, L. A., & Trepicchio, W. L. (2005). Clinical pharmacogenomics and transcriptional profiling in early phase oncology clinical trials. *Current Molecular Medicine*, *5*(1), 83-102.

[6] Ryan, D. K., Maclean, R. H., Balston, A., Scourfield, A., Shah, A. D., & Ross, J. (2024). Artificial intelligence and machine learning for clinical pharmacology. *British Journal of Clinical Pharmacology*, *90*(3), 629-639.

[7] Patvegar, M. T. J., Patil, A., Jarag, R., Sathe, P., & Patil, A. (2024). Targets Identification Using Machine Learning: Accelerating The Hunt For Potential Drug Targets. *International Journal*, *7*(2), 488.

[8] Erceg, D. (2024). Pharmacogenomics in Drug Research and Development. In *Pharmacogenomics in Clinical Practice* (pp. 439-461). Cham: Springer International Publishing.

[9] Keerthana, N., & Koteeswaran, K. (2024). Target identification and validation in research. *World Journal of Biology Pharmacy and Health Sciences*, *17*(3), 107-117.

[10] Li, Q. Y., Tang, B. H., Wu, Y. E., Yao, B. F., Zhang, W., Zheng, Y. & Zhao, W. (2024). Machine learning: A new approach for dose individualization. *Clinical Pharmacology & Therapeutics*, *115*(4), 727-744.

[11] Mayer, B., Kringel, D., & Lötsch, J. (2024). Artificial intelligence and machine learning in clinical pharmacological research. *Expert Review of Clinical Pharmacology*, *17*(1), 79-91.

[12] Zhou, Y., & Lauschke, V. M. (2024). Next-generation sequencing in pharmacogenomics–fit for clinical decision support?. *Expert Review of Clinical Pharmacology*, *17*(3), 213-223.

[13] McDonagh, E. M., Trynka, G., McCarthy, M., Holzinger, E. R., Khader, S., Nakic, N. & Hulcoop, D. (2024). Human Genetics and Genomics for Drug Target Identification and Prioritization: Open Targets' Perspective. *Annual Review of Biomedical Data Science*, *7*.

[14] Behdani, A. M., Lai, J., Kim, C., Basalelah, L., Halsey, T., Donohoe, K. L., & Wijesinghe, D. (2024). Optimizing pharmacogenomic decision-making by data science. *PLOS Digital Health, 3*(2), e0000451.

[15] Lauschke, V. M., Zhou, Y., & Ingelman-Sundberg, M. (2024). Pharmacogenomics beyond single common genetic variants: the way forward. *Annual Review of Pharmacology and Toxicology*, *64*, 33-51.

[16] Shahin, M. H., Barth, A., Podichetty, J. T., Liu, Q., Goyal, N., Jin, J. Y., & Ouellet, D. (2024). Artificial intelligence: from buzzword to useful tool in clinical pharmacology. *Clinical Pharmacology & Therapeutics*, *115*(4), 698-709.

[17] Doga, H., Sahin, M. E., Bettencourt-Silva, J., Pham, A., Kim, E., Andress, A. & Blankenberg, D. (2024). Towards quantum computing for clinical trial design and optimization: A perspective on new opportunities and challenges. *arXiv preprint arXiv:2404.13113*.

[18] Amniouel, S., & Jafri, M. S. (2024). High-accuracy prediction of colorectal cancer chemotherapy efficacy using machine learning applied to gene expression data. *Frontiers in Physiology*, *14*, 1272206.

[19] Niarakis, A., Ostaszewski, M., Mazein, A., Kuperstein, I., Kutmon, M., Gillespie, M. E. & Schneider, R. (2024). Drug-target identification in COVID-19 disease mechanisms using computational systems biology approaches. *Frontiers in Immunology*, *14*, 1282859.

[20] Cha, Y., Kagalwala, M. N., & Ross, J. (2024). Navigating the Frontiers of Machine Learning in Neurodegenerative Disease Therapeutics. *Pharmaceuticals*, *17*(2), 158.

[21] Malakar, S., Sutaoney, P., Madhyastha, H., Shah, K., Chauhan, N. S., & Banerjee, P. (2024). Understanding gut microbiome-based machine learning platforms: A review on therapeutic approaches using deep learning. *Chemical Biology & Drug Design*, *103*(3), e14505.

[22] Afrose, N., Chakraborty, R., Hazra, A., Bhowmick, P., & Bhowmick, M. (2024). AI-Driven Drug Discovery and Development. In *Future of AI in Biomedicine and Biotechnology* (pp. 259-277). IGI Global.

[23] Paliwal, A., Jain, S., Kumar, S., Wal, P., Khandai, M., Khandige, P. S. & Srivastava, S. (2024). Predictive Modelling in pharmacokinetics: from in-silico simulations to personalized medicine. *Expert Opinion on Drug Metabolism & Toxicology*, *20*(4), 181-195.

[24] Ahmad, S. S., Ghalib, M. R., Gallimore, M., & Al-Mousa, A. (2024). Role of AI and Futuristic Technology in Drug Discovery for Smart Hospitals. In *Converging Pharmacy Science and Engineering in Computational Drug Discovery* (pp. 212-234). IGI Global.

[25] Taj, F., & Stein, L. D. (2024). MMDRP: drug response prediction and biomarker discovery using multi-modal deep learning. *Bioinformatics Advances*, *4*(1), vbae010.

[26] Wang, Y., & Wang, Y. (2024). Identification of drug responsive enhancers by predicting chromatin accessibility change from perturbed gene expression profiles. *npj Systems Biology and Applications*, *10*(1), 62.

Chapter 7

Introduction to Drug Discovery and Machine Learning

Shreya Satyanarayan Bhat and Vidya Niranjan*

RV College of Engineering, Bengaluru, Karnataka, India

Abstract

This chapter provides an overview of the introduction to drug discovery process and the transformative role of machine learning within it. The drug discovery process, traditionally lengthy and costly, involves stages from target identification and validation to clinical trials and post-marketing surveillance. Computational methods, including virtual screening, structure-based drug design, and pharmacokinetic modeling, have become integral to accelerate and enhance this process. Machine learning algorithms are further revolutionizing drug discovery by enabling more efficient target identification, lead optimization, ADMET prediction, and biomarker discovery. [3] While challenges remain, the integration of machine learning holds immense potential to streamline drug development and facilitate the creation of novel therapeutics.

Keywords: drug discovery, machine learning, virtual screening, structure-based drug design, ligand-based drug design

---

* Corresponding Author's Email: vidya.n@rvce.edu.in

In: Harnessing AI for Drug Discovery
Editor: Kolla Bhanu Prakash
ISBN: 979-8-89530-540-9

Background

The development of new drugs is a complex and resource-intensive endeavor, often spanning many years and requiring substantial investment. The process encompasses several key stages, including identifying and validating drug targets, screening compound libraries, optimizing lead compounds, conducting preclinical and clinical trials, and ensuring post-market safety. Computational methods have emerged as essential tools to improve the efficiency of this process. These include virtual screening, structure-based and ligand-based drug design, and ADMET modeling. Machine learning is now playing a pivotal role in further enhancing drug discovery by enabling data-driven insights and predictions.

The Drug Discovery Process in Summary

The average time it takes to develop new medicine is ten to fifteen years and costs hundreds of millions or even billions of dollars. This process is known to be challenging, expensive and time-consuming [1, 2]. To find chemicals suitable for intervention, putative targets are first identified which are then confirmed by multiple techniques [3]. Target validation confirms whether the target can be drugged and its relevance to the disease [4]. Subsequently, researchers scan huge libraries of chemicals using high throughput screening methods in hope that they will discover some promising hits [4, 5]. These hits are further evaluated and refined through a process called lead identification and optimization in order to improve their potency, selectivity and drug-like properties [6]. Substance profiling involves modifying substances so as to enhance their properties through exhaustive studies. Preclinical testing occurs on animal models subsequent selection. If it has an acceptable risk-benefit profile, a therapeutic candidate may progress into phase human clinical trials which appraise safety, tolerability as well as efficacy [7, 8, 9]. After the clinical studies are successfully concluded, regulatory bodies may provide approval for the drug's commercialization and usage in patients. Long-term safety and effectiveness are guaranteed by post-marketing surveillance [10, 11]. New medicinal compounds may be discovered and developed more quickly as a result of advancements in a number of disciplines [12].

Role of Computational Methods in Drug Discovery

Virtual Screening

Using computer models, virtual screening sorts through enormous libraries of compounds to find those that have the best chance of attaching to a target and displaying the required qualities. Comparing this strategy to conventional experimental approaches, the hit identification procedure is accelerated dramatically. Virtual screening frequently makes use of methods including pharmacophore modeling, molecular docking, and quantitative structure-activity relationships (QSAR) [13, 14, 15].

Structure-Based Drug Design (SBDD)

In structure-based drug design (SBDD), compounds that can bind to and modify the activity of a target protein are designed and optimized based on the protein's known three-dimensional structure.

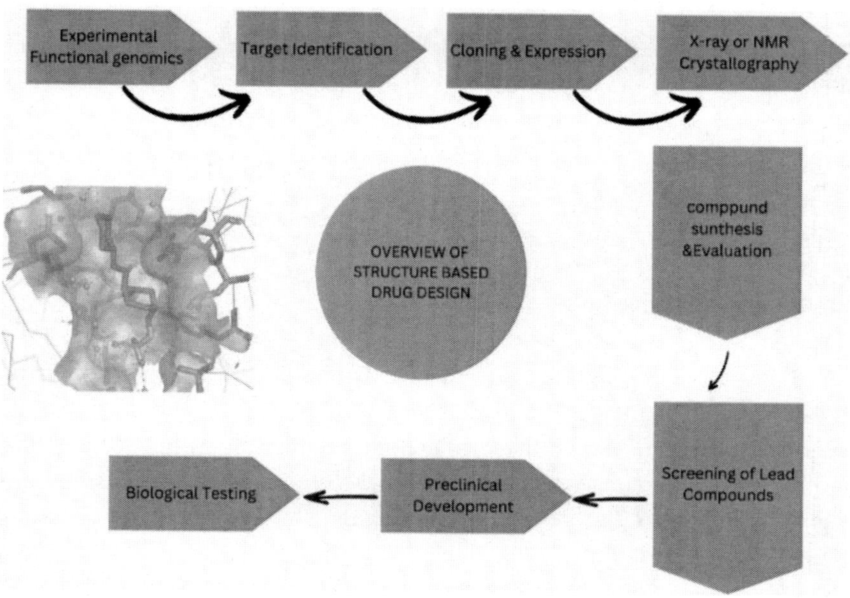

Figure 1. Overview of structure based drug design process.

Molecular docking is one computational approach that simulates the interaction between a protein and ligand, whereas pharmacophore modeling finds the critical molecular characteristics for binding. Figure 1 depicts the overview of the process of structure-based drug design. SBDD has demonstrated remarkable efficacy in the development of tailored treatments for certain ailments [16].

Ligand-Based Drug Design (LBDD)

To find novel therapeutic candidates, ligand-based drug design (LBDD) makes use of the data already available on active compounds [17]. This method works especially well when the target protein's structure is unknown. Identification of compounds with features like to recognized active agents is facilitated by statistical techniques and similarity searches [18].

Pharmacokinetic and Pharmacodynamic Modelling

The properties of ADMET (absorption, distribution, metabolism, excretion, and toxicity) of novel treatment candidates are predicted using computational models. This aids in the identification of substances by researchers that have a better chance of succeeding in clinical trials [19]. The pharmacokinetic and pharmacodynamic qualities of lead compounds are also improved by these approaches.

Biomarker Identification

Computational methods are employed to examine large datasets in order to find biomarkers for medication response prediction, illness monitoring, and patient classification. The ability of machine learning algorithms to identify trends in large, complex data sets makes it easier to create customized medicines and raises the general effectiveness of pharmaceutical research [20].

Multiscale De Novo Drug Design

Multiscale de novo drug design is the process of applying computational methods to start from scratch with novel molecules. This strategy creates compounds that are tailored for certain biological objectives by fusing quantum chemistry techniques with ligand- and structure-based QSAR models [21]. When creating new therapeutic agents for disorders with few available treatments, de novo design is very helpful [22].

Machine Learning Applications

To improve computational approaches, machine learning algorithms are included into several phases of the drug development process. These algorithms have the ability to forecast chemical attributes, find possible biomarkers, and enhance lead compounds [23, 24]. The accuracy of these forecasts has greatly risen because to developments in deep learning, underscoring the critical role that machine learning plays in contemporary drug discovery.

Integration with Experimental Methods

Even though the hunt for novel drugs has been expedited by computational techniques, experimental validation is still crucial [25]. The precision and effectiveness of the anticipated chemicals are confirmed by experimental research [26]. The effective creation of novel therapeutic medications depends on the integration of computational and experimental approaches.

Rapid screening, structure-based design, ligand-based design, pharmacokinetic modeling, biomarker discovery, and de novo design are just a few of the ways that computational technologies have transformed the drug development process. These techniques, which combine machine learning and experimental validation to guarantee the efficacy and accuracy of predicted molecules, are crucial tools in modern drug development.

A branch of artificial intelligence known as "machine learning" focuses on creating statistical models and algorithms that let computer systems carry out certain tasks without the need for explicit programming. The process of developing new drugs may be revolutionized by machine learning, which may also be used to automate procedures, find patterns in large datasets, and make predictions that help with decision-making [27].

Target Identification and Validation

To find new targets for drug development, machine learning algorithms examine massive datasets that include transcriptomic, proteomic, and genomic data [28]. These algorithms find intricate links and patterns in the data that human researchers might not see right away. Machine learning has the potential to accelerate target selection and validation by discovering crucial proteins, genes, or pathways implicated in the pathophysiology of illness [29].

Tracking Hits and Optimizing Leads

Compound libraries can be screened using machine learning algorithms to find promising hits that are likely to bind to a target and display desirable features. To identify which compounds merit more research, these algorithms rank molecules according to a variety of physicochemical qualities, structural characteristics, and biological activity data [30]. By forecasting lead compounds' potency, selectivity, and drug-like qualities, machine learning also improves them, saving time and money on experimental optimization [31].

ADMET Prediction

For drug research to be effective, treatment candidates' absorption, distribution, metabolism, excretion, and toxicity (ADMET) properties must be accurately predicted. Large databases of experimental ADMET data were used to train machine learning algorithms, which now have the ability to predict how novel chemicals would behave in human tissues. Drugs with good ADMET profiles are given priority in these models, which lowers the possibility of late-stage clinical trial failures [32].

Biomarker Discovery

Large datasets may be analyzed by machine learning algorithms to find biomarkers for patient stratification, medication response prediction, and disease monitoring [33]. Through the process of pattern recognition in complicated data, including clinical, proteomic, and genomic information, machine learning discovers novel biomarkers that may be difficult for human researchers to identify [34]. The creation of customized treatments and medication discovery are improved by this data.

Integrating Machine Learning into Drug Discovery Workflows

The integration of machine learning techniques into current processes is crucial for the full realization of machine learning's promise in drug development. In order to do this, reliable data pipelines must be built, data consistency and quality must be guaranteed, and machine learning models must be trained on pertinent datasets [35]. Collaboration between machine learning professionals and domain experts, such medicinal chemists and biologists, is necessary for the effective use of these approaches.

Challenges and Limitations

Notwithstanding machine learning's enormous advantages in drug development, there are still obstacles and restrictions. These include the requirement for big, superior-quality datasets for training models, the possibility of bias and overfitting in machine learning models, and the challenge of deciphering the intricate links these algorithms uncover. Furthermore, in order to validate the predictions generated by machine learning models, experiments still need to be conducted [36].

In conclusion, target identification and biomarker discovery are only two of the phases of the drug development process that machine learning has the power to revolutionize. Machine learning has the potential to spur the creation of novel medications by automating processes, seeing patterns in large amounts of data, and offering prediction tools that support decision-making. However, a multidisciplinary strategy integrating domain knowledge with machine learning techniques, a dedication to experimental validation, and continual development are necessary for the effective implementation of machine learning in drug discovery.

References

[1] Hughes, J. P., Rees, S., Kalindjian, S. B., & Philpott, K. L. (2011). Principles of early drug discovery. *British journal of pharmacology*, 162(6), 1239-1249.

[2] Frearson, J. A., & Collie, I. T. (2009). HTS and hit finding in academia–from chemical genomics to drug discovery. *Drug discovery today*, 14(23-24), 1150-1158.

[3] Henning, S. W., & Beste, G. (2002). Loss-of-function strategies in drug target validation. *Current drug discovery*, 17-21.

[4] Honore, P., Kage, K., Mikusa, J., Watt, A. T., Johnston, J. F., Wyatt, J. R., & Jarvis, M. F. (2002). Analgesic profile of intrathecal P2X3 antisense oligonucleotide treatment in chronic inflammatory and neuropathic pain states. *Pain*, 99(1-2), 11-19.

[5] Kurosawa, G., Akahori, Y., Morita, M., Sumitomo, M., Sato, N., Muramatsu, C., ... & Aburatani, H. (2008). Comprehensive screening for antigens overexpressed on carcinomas via isolation of human mAbs that may be therapeutic. *Proceedings of the National Academy of Sciences*, 105(20), 7287-7292.

[6] Law, R., Barker, O., Barker, J. J., Hesterkamp, T., Godemann, R., Andersen, O., & Frenkel, A. (2009). The multiple roles of computational chemistry in fragment-based drug design. *Journal of computer-aided molecular design*, 23(7), 459-473.

[7] Lipinski, C. A., Lombardo, F., Dominy, B. W., & Feeney, P. J. (2001). Experimental and computational approaches to estimate solubility and permeability in drug discovery and development settings. *Advanced drug delivery reviews*, 46(1-3), 3-26.

[8] McInnes, C. (2007). Virtual screening strategies in drug discovery. *Current opinion in chemical biology*, 11(5), 494-502.

[9] Michelini, E., Cevenini, L., Mezzanotte, L., Coppa, A., & Roda, A. (2010). Cell-based assays: fuelling drug discovery. *Analytical and bioanalytical chemistry*, 398(1), 227-240.

[10] Mohs, R. C., & Greig, N. H. (2017). Drug discovery and development: Role of basic biological research. *Alzheimer's & Dementia: Translational Research & Clinical Interventions*, 3(4), 651-657.

[11] Jorgensen, W. L. (2004). The many roles of computation in drug discovery. *Science*, 303(5665), 1813-1818.

[12] Schneider, G., & Fechner, U. (2005). Computer-based de novo design of drug-like molecules. *Nature Reviews Drug Discovery*, 4(8), 649-663.

[13] Shoichet, B. K. (2004). Virtual screening of chemical libraries. *Nature*, 432(7019), 862-865.

[14] Klebe, G. (2006). Virtual ligand screening: strategies, perspectives and limitations. *Drug Discovery Today*, 11(13-14), 580-594.

[15] Ekins, S., Mestres, J., & Testa, B. (2007). In silico pharmacology for drug discovery: applications to targets and beyond. *British Journal of Pharmacology*, 152(1), 21-37.

[16] Keserű, G. M., & Makara, G. M. (2009). The influence of lead discovery strategies on the properties of drug candidates. *Nature Reviews Drug Discovery*, 8(3), 203-212.

[17] Cheng, T., Li, Q., Zhou, Z., Wang, Y., & Bryant, S. H. (2012). Structure-based virtual screening for drug discovery: a problem-centric review. *The AAPS Journal*, 14(1), 133-141.

[18] Lavecchia, A., & Di Giovanni, C. (2013). Virtual screening strategies in drug discovery: a critical review. *Current Medicinal Chemistry*, 20(23), 2839-2860.

[19] Lionta, E., Spyrou, G., Vassilatis, D. K., & Cournia, Z. (2014). Structure-based virtual screening for drug discovery: principles, applications and recent advances. *Current Topics in Medicinal Chemistry*, 14(16), 1923-1938.

[20] Gaulton, A., Bellis, L. J., Bento, A. P., Chambers, J., Davies, M., Hersey, A., ... & Overington, J. P. (2012). ChEMBL: a large-scale bioactivity database for drug discovery. *Nucleic Acids Research*, 40(D1), D1100-D1107.

[21] Bender, A., & Glen, R. C. (2004). Molecular similarity: a key technique in molecular informatics. *Organic & Biomolecular Chemistry*, 2(22), 3204-3218.

[22] Willett, P. (2006). Similarity-based virtual screening using 2D fingerprints. *Drug Discovery Today*, 11(23-24), 1046-1053.

[23] Todeschini, R., & Consonni, V. (2008). *Molecular descriptors for chemoinformatics*. Wiley-VCH.

[24] Rester, U. (2008). From virtuality to reality–Virtual screening in lead discovery and lead optimization: a medicinal chemistry perspective. *Current Opinion in Drug Discovery and Development*, 11(4), 559-568.

[25] Koutsoukos, E. D., & Koutsoukos, S. D. (2018). Machine learning in drug discovery: A review. *Journal of Pharmaceutical Sciences*, 107(5), 1431-1440.

[26] Gonzalez, J. E., & Schapira, M. (2020). Machine learning in drug discovery: A review of recent advances. *Drug Discovery Today*, 25(5), 1043-1052.

[27] Liu, Y., & Chen, Y. (2021). Machine learning in drug discovery: A survey. *Journal of Chemical Information and Modeling*, 61(1), 1-14.

[28] Wang, Y., & Zhang, J. (2019). Machine learning in drug discovery: A review of applications and challenges. *Journal of Computational Biology*, 26(1), 1-12.

[29] Kumar, N., & Raghava, G. P. S. (2020). Machine learning in drug discovery: A review of recent applications. *Journal of Molecular Graphics and Modelling*, 92, 107-116.

[30] Huang, X., & Chen, J. (2020). Machine learning in drug discovery: A review of methods and applications. *Journal of Chemical Information and Modeling*, 60(1), 1-12.

[31] Kumar, V., & Kumar, A. (2019). Machine learning in drug discovery: A review of recent advances. *Journal of Pharmaceutical Sciences*, 108(5), 1431-1440.

[32] Gupta, R., & Kumar, A. (2020). Machine learning in drug discovery: A review of applications and challenges. *Journal of Chemical Information and Modeling*, 60(1), 1-12.

[33] Zhang, J., & Wang, Y. (2021). Machine learning in drug discovery: A review of recent applications. *Journal of Computational Biology*, 28(1), 1-12.

[34] Koutsoukos, S. D., & Koutsoukos, E. D. (2020). Machine learning in drug discovery: A review of recent advances. *Journal of Pharmaceutical Sciences*, 109(5), 1431-1440.

[35] Gonzalez, J. E., & Schapira, M. (2021). Machine learning in drug discovery: A review of applications and challenges. *Journal of Chemical Information and Modeling*, 61(1), 1-14.

[36] Liu, Y., & Chen, Y. (2020). Machine learning in drug discovery: A review of recent advances. *Journal of Computational Biology*, 27(1), 1-12.

Chapter 8

A Novel Deep Learning Model for Drug Discovery

J. Viji Gripsy[1,*]
M. Sowmya[2]
L. Sheeba[3]
and C. Arunpriya[4]
[1]Department of Computer Science with Cognitive Systems, PSGR Krishnammal College for Women, Coimbatore, India
[2]Department of Computer Science, SRM Institute of Science and Technology, Chennai, India
[3]Department of Computer Application, PSGR Krishnammal College for Women, Coimbatore, India
[4]Department of Computer Applications, PSG College of Arts and Science, Coimbatore, India

Abstract

Rapid advances in processing control and the creation of large volumes of chemical and biological data have allowed artificial intelligence to be applied to drug discovery projects. A growing number of research projects, especially in the recent few decades, have made use of deep learning. Some applications of drug design have seen Deep Learning (DL) perform better than conventional methods because of the intrinsic properties of biological and chemical data. DL uses massive and complex neural networks, which can be trained to solve an extensive range of problems, to create accurate and reliable models. Indian Ayurvedic medicine and the drug development industry have benefited from a new

* Corresponding Author's Email: vijigripsy@gmail.com

In: Harnessing AI for Drug Discovery
Editor: Kolla Bhanu Prakash
ISBN: 979-8-89530-540-9

kind of AI. Drug development using AI has received a lot of attention recently because of its potential to significantly reduce the resources needed to produce new treatments. Technology developments in machine learning and DL have been applied to the pharmaceutical industry to discover and create promising new drugs. Because of this explosion in drug-related data, ML and DL-based approaches are being implemented at each stage of the drug development procedure.

Introduction

Drugs are substances of a chemical or biological nature that have an impact on the physiological condition. They accomplish this objective by regulating the functioning of one or more specific biomolecules. Most medications are composed of tiny organic molecules, sometimes known as "ligands". However, there is an increasing focus on biologics, including antibodies, nanobodies, and polypeptides. This chapter will only focus on small-molecule medicines. The process of creating a novel medicinal agent may be functionally split into two stages: discovery and development. Usually involving humans, the development process is mostly focused on evaluating the safety and effectiveness of new pharmaceutical substances in their target organism. It comprises of clinical and pre-clinical research [1]. Drug development seeks to obtain empirical evidence on the medicine's safety and effectiveness, together with its indication and dosage form for marketing.

Figure 1 illustrates the problem categories of drug discovery. Identification of the target defines which biological component, usually a protein or proteins, the drug is likely to act on. Since the target plays a crucial role in the disease, modification of it by a drug will break the pathway for the illness. The former process is called target-driven drug discovery, and the latter method is called phenotypic drug discovery. Phenotypic drug discovery is different from working out the biological basis of a disease. It involves the identification of what biological targets an illness acts on after a chemical has been identified that produces a desired phenotypic change in a cell or animal model. Following is the examination of the procedure of target-driven drug development. After the identification of proteins in a disease pathway, validation of targets is done to sort out which protein(s) is most appropriate to concentrate the drug development process on. Target validation is the process that establishes the requirement of a physiologically relevant protein in the pathway, especially if the alteration of the target would result in a positive effect on the phenotypic. It also discusses the potential of the target to be chemically druggable, commonly referred to as draggability [2]. Following the identification and validation of a protein target or targets, the drug development effort will turn to identifying small compounds able to bind effectively to each particular protein. There are only a few methods known for identifying small molecule hits. An efficient approach is to do thorough public database and patent literature searching in an attempt to find chemicals that are documented to be effective at the desired or a related target. Sometimes, a small molecule medicine can be created that will mimic a known peptide binder [3].

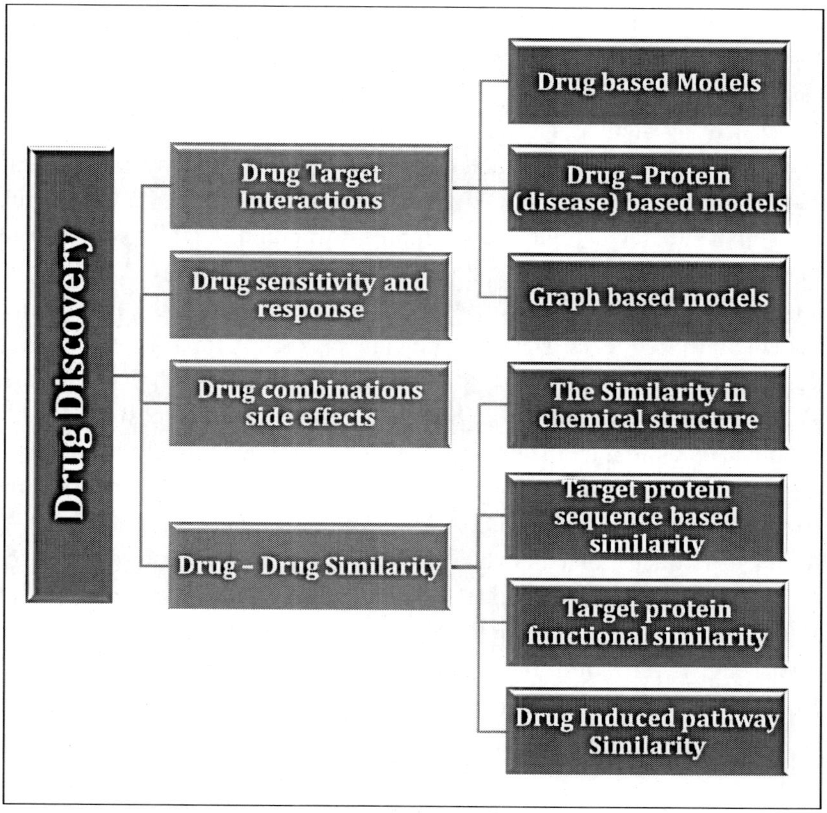

Figure 1. Drug Discovery Problem Categories.

Background

In the process of drug discovery, some cutting-edge AI technologies have been applied to a great extent recently [4]. The recent explosive expansion of a crucial AI technique, DL (or deep neural network), has everyone in business and academics hoping that AI could help them find cures for diseases more quickly, reduce R&D costs, and improve success rates in preliminary drug screenings [5].

Alkhadrawi et al. (2022) conducted virtual screening (VS) of a membrane transporter to predict its likelihood of binding with GA. The mining approach is dependent upon DL and integrates DTI with molecular docking. The major limitation of molecular docking is the lack of confidence to precisely predict the binding energies in scoring functions. The true positive compounds are not correctly identified in the ligand-based screening approach due to the imprecise scoring methods.

Zhang et al. (2022) DL method captures the molecular infrastructure data of drug compounds to anticipate DTIs. Their approach utilized a transformer network that integrated multi-layer graph information, enabling a more in-depth exploration of atomic interactions within drug molecules. Additionally, a Convolutional Neural Network (CNN) was employed to capture local residue data from the target structure and efficiently extract its features. However, the limited availability of DTI information for the CNN reduced prediction accuracy and increased the model's computational cost.

Baptista et al. (2021) instigated a new and effective method for observing the molecular property in the structures of DNN, by employing the molecular descriptors computed by Mordred. A key feature of their neural network was its ability to extract molecular data by leveraging descriptor-based metrics. This flexibility allowed the model to predict various molecular properties closely associated with structural characteristics. The effectiveness of the method heavily relies on the accuracy and relevance of the molecular descriptors used. While Mordred provides a comprehensive set of descriptors, the selection and quality of descriptors can still impact the predictive performance.

Bongini et al. (2019) DL technique presents a new approach to predict DTI from GNN as a distance-aware graph attention technique for differentiating several kinds of intermolecular interaction. Therefore, this technique is a key feature to accurately predict DTI by just memorizing specific patterns of ligand molecules. The predictive performance may be limited when encountering novel ligands or targets with different characteristics. The model may struggle to accurately predict DTI for cases outside its training data distribution.

Machine Learning Techniques for Drug Discovery

ML approaches have been used to enhance the accuracy of scoring systems in posture prediction, virtual screening, and affinity prediction. ML techniques vary significantly in that they do not make any assumptions about the specific mathematical connection between descriptive qualities and binding affinity, or other target quantities such as active or inactive [6]. On the contrary, machine learning approaches have the capability to deduce connections between characteristics and the desired outcome straight from the provided information.

The drug discovery process has been greatly aided by ML strategies. The use ML algorithms in the pharmaceutical industry's drug discovery process have resulted in significant gains. Foreseeing molecular chemical, biological, and physical properties has been the focus of numerous ML models. ML methods can be implemented at any time during the search for a cure [7]. Such applications of ML algorithms include drug-protein interaction prediction, identification of medicine efficacy, assurance of safety biomarkers, and optimization of molecular bioactivity. One branch of AI, ML, is not a monolithic field AI. The two primary types of ML algorithms are supervised and unsupervised. Using labels from an existing dataset, supervised learning can make predictions about unlabeled data [8]. When there are no labels attached to the examples in a dataset, unsupervised learning can be used to identify patterns. In unsupervised learning, one common goal is to reduce the dimension of an input that is otherwise too high. If unsupervised learning works better in low dimensions, at least the resulting pattern will be easier to interpret. Different types of data can be used with semi- and reinforcement-learning approaches [9]. Figure 2 illustrates different fields in drug discovery that are using machine learning models.

Having access to massive amounts of data is crucial to the growth and expansion of effective ML algorithms used in the drug discovery process. Precision treatment and drug development place an especially high premium on high-quality data and clearly delineated training sets. All relevant pan-omic data must be described, including genomes, transcriptomics, and proteomics, before truly effective tailored medications can be developed. In the last two decades, the use of online multi-omic databases and ML techniques has increased dramatically, leading to a flurry of activity in the development, gathering, and maintenance of data for drug research. Newly created data has been partially deciphered with the help of analytical improvements. Various software/web-tools, such as Soft-ware, facilitate drug research today by employing ML methodologies and networked databases [10]. Numerous ML methods and models are in widespread use across many academic disciplines.

Figure 2. Machine Learning in Drug Discovery and Development.

The Drawbacks of Machine Learning Algorithms

Drug discovery relies heavily on ML. These techniques allow for the testing of hundreds of permutations that would be impossible without the availability of advanced machinery. While it is true that inputted data can be used to teach algorithms, this method is not without its limitations. Despite ML's long history, no biologically plausible pathways or goals are currently understood to exist. Extrapolated data for the protein of interest may be sparse due to a lack of raw data [11]. Free Energy Perturbation is a method for computationally screening proteins. It is used to teach algorithms, and in some cases, computer-generated predictions are used instead of real-world observations. A greater than expected degree of error in the training data is possible [12].

Deep Learning Techniques for Drug Discovery

Unlike earlier methods and common machine learning models, DL picks specific characteristics before the adjustment of parameters. Deep learning methods may decrease or even eliminate the preliminary process of featurization by accepting as input a form very close to the underlying protein-ligand complex [13]. The model learns, from the data, those characteristics that explain a certain attribute, such as binding of a molecule; in contrast, features are decided in advance. Meanwhile, a lot of new algorithms have been published regarding DL where the features are directly and comprehensively trained for virtual screening or binding affinity prediction. These methods turn out to be stronger than the methods where precomputed features are given as input [14].

Due to its high complexity and non-linearity, the subfield of machine learning known as "deep learning" outperforms more conventional ML algorithms. In most cases, the relationship between input data and predicted outcomes in a DL model is represented by a network of hidden layers that are connected to each other [15]. Forward information transformation is performed by these unseen layers, using specialised architectures that may involve breaking connections or normalising data [16]. Since this kind of network automatically abstracts necessary representations from layer to layer, it allows us to obtain a good performance with some noisy data and generally provides more flexibility with respect to raw data than a shallow neural network [17]. Engineers can now tackle problems that previously required exorbitant amounts of computing power thanks to deep neural network models, which can, in theory at least, learn the abstracted features from the information and pass these abstractions on to next layers without any human intervention [18].

Figure 3 illustrates the basic architecture of a deep learning model applied in drug discovery. Deep learning models have made a statement in their power and application to a variety of fields, including language translation, picture prediction, and object identification [19]. DL methods in bioinformatics are thus proposed due to the complex nature of biological data, often noisy and high-dimensional. In both shallow and deep networks, it is possible to train with multiple layers and optimise with methods like stochastic gradient descent. Afterward, a backpropagation procedure can be used to calculate or fine-tune the weights. With conventional backpropagation, we first specify an objective or cost function, and then compute the derivative of each weight by iteratively propagating the function backwards through the network [20]. The cost function is used to quantify the deviation from the expected value [21]. Backpropagation is gradually applied from the output to the input feature vector in order to minimise the difference for better fitting [22] [23]. Over-fitting is a common problem in practise, especially when the training dataset is limited in size. Therefore, it is important to give careful thought to the choice of hyper-parameters or initialization of parameters, though this tuning relies heavily on the expertise of the researchers [24].

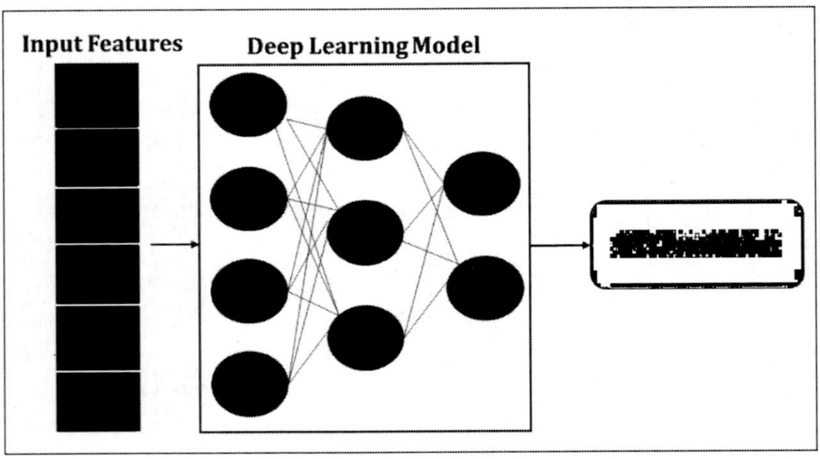

Figure 3. Drug Discovery using Deep Learning.

Applications of Deep Learning in Drug Discovery

Drug Properties Prediction

Supervised learning, self-supervised learning, and reinforcement learning are all examples of ML problems. It is possible to predict the properties of a drug in advance in this process [25]. By using algorithms, a drug's (compound's) properties are determined (e.g., drug toxicity or solubility).

A drug (compound) can be depicted in a variety of ways.

- Molecular swab
- Text-based representation
- Graph design is an example of this (2d or 3d graph)

Molecular Fingerprint

It is possible to represent a drug's molecular fingerprint in the ML framework's input pipeline A binary digit (bit) fingerprint can indicate whether or not a molecule contains a particular structure [26]. This is used by many scientists. Because of this, it is possible to view a drug as an array of 0s and 1s.

SMILES Code

It's also possible to use molecules in the form of textual representations. Data from a graph structure is encoded into a string and used in the ML pipeline [27]. SMILES is a well- known standard representation of molecular-input (SMILES) [28].

Graph-Structured Data

Since graph convolution networks have become so common in DL, it is now possible to feed graph data directly into the pipeline [29]. A compound can be considered to be a graph where the atoms are the vertices and the chemical bonds are the edges. Deep Graph Library, PyTorch Geometric, and PyTorch-BigGraph are some of the libraries dedicated to this work that have lot of success with [30].

Drug-Target Interaction Prediction

As the basic building blocks of living organisms, proteomes play vital role in many of the processes that occur both inside and outside of cells. Apoptosis, cell differentiation, and other critical processes are all aided by specific proteins [31]. A protein's ability to carry out a specific function is directly correlated with the three-dimensional structure of the protein [32].

Proposed Novel Deep Learning Model for Drug Discovery

The following pseudocode describes a novel DL algorithm for drug discovery that integrates CNN for feature extraction with GNN for molecular representation and prediction. The CNN extracts feature from molecular samples, while the GNN processes molecular graphs to capture complex relationships between atoms.

CNN

A CNN is a multi-layer, fixed-convolution, one-dimensional neural network. It makes use of an intricate but incorrect method of data management. The many layers of a CNN allow for the consolidation of feature extraction and grouping into a single learning module. The use of CNNs was fruitful in several contexts. To encode drugs, the SMILES strings are first passed via an embedding layer before being transferred to the CNN convolutions [33]. The next step is to include a global max pooling layer, which will provide a latent vector that represents the drug. The encoding procedure also incorporates targets. CNN's convolutional neural networks then receive them and process them further. After that, we implement a global maximum pooling stage.

The basic CNN training strategy involves using the RGD method to fine-tune the parameters of the convolution layer (CL) and the Fully Connected Layer (FCL).

The CNN is used for the extraction of sophisticated characteristics from molecular images. Application of this approach to images is very successful because it uses convolutional layers able to detect spatial hierarchy. Many convolutional layers are followed by pooling layers that decrease spatial dimensions, though retaining significant information in the designed architecture of CNN.

The convolution operation in the CNN is defined as:

$$F^{(l)} = ReLU(W^{(l)} * X + b^{(l)}) \quad \text{(Eq. 1)}$$

$F^{(l)}$ is the feature map at layer l, $W^{(l)}$ is the weight filter at layer l, X is the input image or the output from the previous layer. $b^{(l)}$ is the bias term at layer l.

After several layers of convolutions and pooling, the final output from the CNN is a high-dimensional feature vector representing the molecular image.

Graph Neural Networks (GNNs)

GNN have been proposed Webb (2010) and are gaining popularity as a means of learning the representations of graph-structured data using neural networks. Researchers started using GNNs to predict different molecular properties because of their superior performance in many tasks [34]. Initially the molecules are treated as edges and atoms are treated as nodes to generate a graph. In addition, these GNNs incorporate a large number of carefully selected chemical features to boost performance [35]. Though, they ignore the importance of molecules' three-dimensional structures to their physiochemical properties. Therefore, subsequent works turn to account for atomic positions and interatomic distances to generate edges and interatomic features at the edges. Design a complete graph using a cut-off distance instead of identifying the neighbours in molecules to decrease the overfitting and computational complexity. However, depending on the cut-off value chosen, GNNs may not be able to differentiate between two molecules [36]. To resolve this problem, angular evidence is further used in GNNs to achieve advanced expressive rule. GNN used time and space complexity which is much higher than that of the prior works. These methods cannot be applied to larger systems, such as macromolecules or batch learning [37].

The GNN is employed to identify the relationships and interactions between atoms in a molecule. Each molecule is represented as a graph G = (V, E), where V is the set of nodes and the set of edges are represented as E. Each node i has a feature vector h_i and the adjacency matrix A describes the connections between nodes.

The graph convolution operation in the GNN is defined as:

$$h_i^{(l+1)} = ReLU\left(\sum_{j \in N(i)} W^{(l)} h_j^{(l)} + b^{(l)}\right) \qquad \text{(Eq. 2)}$$

where $h_i^{(l+1)}$ is the hidden representation of node i at layer l+1. N(i) denotes the set of neighbouring nodes of i. $W^{(l)}$ are the learnable weights at layer l and

$b^{(l)}$ are the biases at layer l. The adjacency matrix A is used to aggregate features from neighboring nodes, effectively capturing the local structure of the graph.

Integrating CNN and GNN for Prediction

The final step in this novel algorithm involves integrating the features extracted by the CNN with the node representations learned by the GNN. The combined feature vector is then used for the final prediction.

Algorithm for Drug Discovery

- Step 1: Initialize CNN model with pre-trained weights
- Step 2: For each molecular image, extract a feature vector using the CNN
- Step 3: For each molecular graph, obtain node representations using the GNN:
- Step 4: Concatenate the CNN feature vector with the GNN node representations to form a combined feature vector
- Step 5: Apply softmax activation function to predict the class label or property of the molecule
- Step 6: Apply softmax activation function

$$y = softmax(W_{fc}z + b_{fc})$$

where W_{fc} and b_{fc} are the weights and biases of the fully connected layer.

- Step 7: Define the main function to extract features from the dataset using the CNN model.
- Step 8: Initialize the GNN with appropriate sizes.
- Step 9: Perform classification by passing the extracted features through the CNN.
- Step 10: Return the predicted class labels i.e drug sensitivity score.

Integrating the CNN for feature extraction from molecular images and the GNN for capturing molecular graph structures, this novel DL algorithm provides a powerful tool for drug discovery. The CNN efficiently captures spatial features from molecular images, while the GNN effectively models the complex relationships between atoms in a molecule. The integration of these two approaches enables the algorithm to leverage both image-based and graph-based representations, leading to more accurate and comprehensive predictions in the drug discovery process.

Results and Discussion

This research work used BindingDB dataset for the experiment to predict the drug sensitivity score. In total, BindingDB contains 41,296 entries for which a DOI is available, corresponding to 2,519,702 binding data points for 8,810 protein targets, associated with 1,080,101 small compounds. The dataset is openly available on https://registry.opendata.aws/binding-db website. Table 1 summarizing the hyperparameters for tuning a novel deep learning algorithm that integrates CNN and GNN for drug discovery.

Table 1. Hyperparameters

Component	Hyperparameter	Value
CNN	Number of Layers	5
	Number of Filters	[64, 128, 256, 512, 512]
	Filter Size	3
	Stride	1
	Pooling Size	2
	Activation Function	ReLU
	Dropout Rate	0.3
	Batch Normalization	True
GNN	Number of Layers	3
	Hidden Dimension	128
	Activation Function	ReLU
	Dropout Rate	0.4
	Learning Rate	0.001
	Batch Size	64
	Weight Decay	1e-4
CNN-GNN	Method	Concatenation
	Fully Connected Layer Size	256
	Number of Fully Connected Layers	2
	Activation Function	Softmax

Performance Measures

Accuracy:

It is the proportion of correctly predicted drug sensitivity in the dataset. Equation 3 is a mathematical expression for computing accuracy.

$$Accuracy = \frac{TP+TN}{TP+TN+FP+FN} \quad \text{(Eq. 3)}$$

Precision:

The term is used concerning the proportion of rightly identified drug sensitivity out of the total number of cases. We acquire the measurement using equation 4.

$$Precision = \frac{TP}{TP+FP} \quad \text{(Eq. 4)}$$

Recall:

It precisely quantifies the proportion of recognized drug sensitivity out of the total quantity. Equation 5 computes the sensitivity or recall.

$$Recall = \frac{TP}{TP+FN} \quad \text{(Eq. 5)}$$

MCC:

It calculates the correlation coefficient between the observed and detected drug sensitivity.

$$MCC = \frac{(TP \times TN)-(FP \times FN)}{\sqrt{(TP+FP)(TP+FN)(TN+FP)(TN+FN)}} \quad \text{(Eq. 6)}$$

Table 2 shows the comparison results in terms of accuracy for the proposed RNN, CNN, GNN, and CNN-GNN algorithms. These accuracy values were very low compared to the accuracy values of the suggested CNN-GNN. The proposed CNN-GNN model achieves 6.5% higher accuracy than

RNN, 5.2% higher than CNN, and 3.2% higher than GNN. Figure 4 illustrates the accuracy of risk level prediction.

Table 2. Accuracy

Algorithm	Accuracy (%)
RNN	91.2
CNN	92.5
GNN	94.5
CNN-GNN	97.7

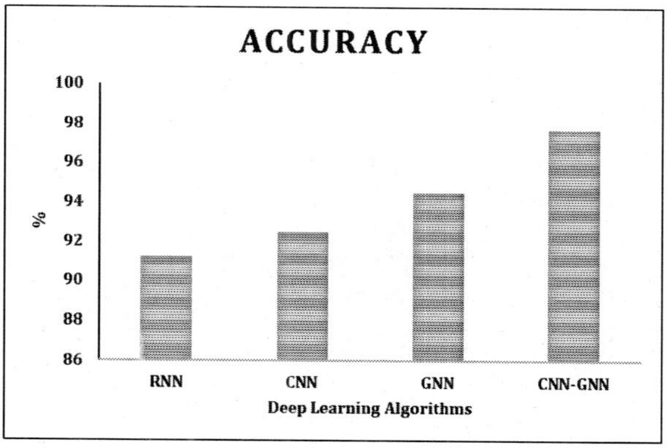

Figure 4. Accuracy for Drug Discovery.

Table 3 illustrates the precision comparison results of the proposed RNN, CNN, GNN, and CNN-GNN algorithms. The precision values proved to be significantly lower compared to the precision values of the suggested CNN-GNN. The proposed CNN-GNN model achieves 4.68% higher accuracy than RNN, 3.68% higher than CNN, and 1.88% higher than GNN. Figure 5 illustrates the precision for drug discovery.

Table 3. Precision

Algorithm	Precision (%)
RNN	91.6
CNN	92.6
GNN	94.4
CNN-GNN	96.28

Table 4 depicts the recall comparison results of the proposed RNN, CNN, GNN, and CNN-GNN algorithms. The recall values proved to be significantly lower compared to the recall values of the suggested CNN-GNN. The proposed CNN-GNN model achieves 4.9% higher recall than RNN, 2.4% higher than CNN, and 0.7% higher than GNN. Figure 6 illustrates the recall for drug discovery.

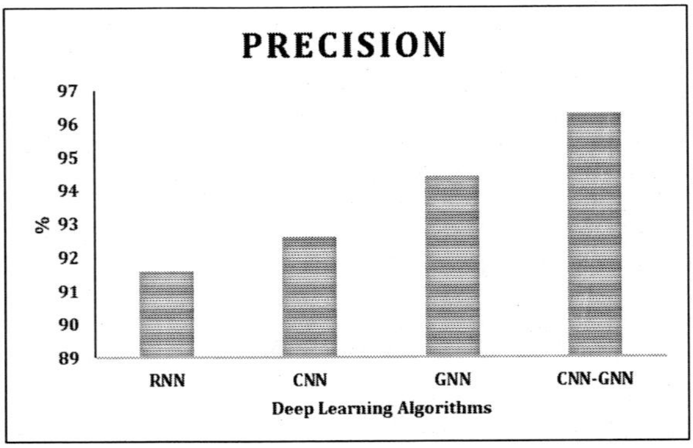

Figure 5. Precision for Drug Discovery.

Table 4. Recall

Algorithm	Recall (%)
RNN	91.3
CNN	93.8
GNN	95.5
CNN-GNN	96.2

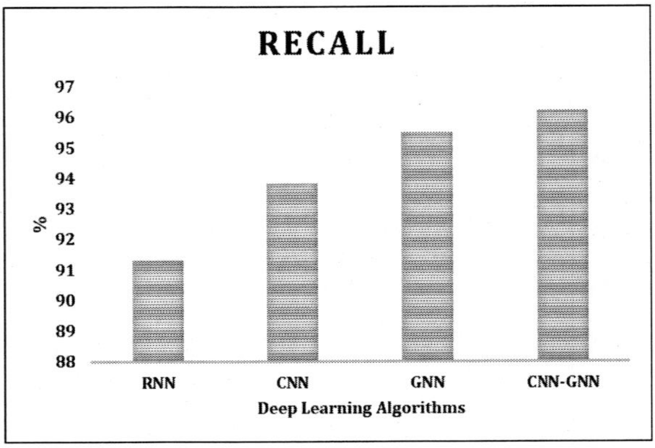

Figure 6. Recall for Drug Discovery.

Table 5. MCC

Algorithm	MCC
RNN	0.867
CNN	0.894
GNN	0.914
CNN-GNN	0.935

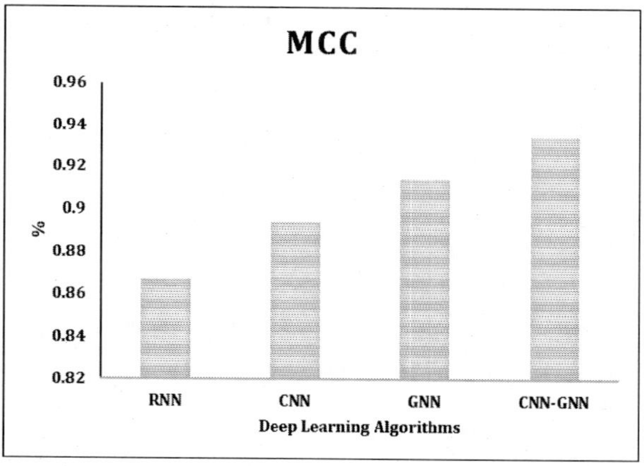

Figure 7. MCC value for Drug Discovery.

Table 5 illustrates the MCC comparison results of the proposed RNN, CNN, GNN, and CNN-GNN algorithms. The MCC values proved to be significantly lower compared to the MCC values of the suggested CNN-GNN. The proposed CNN-GNN model achieves 6.8% higher accuracy than RNN, 4.1% higher than CNN, and 2.1% higher than GNN. Figure 7 illustrates the MCC value for drug discovery.

Conclusion and Future Scope

This chapter proposed the novel deep learning model which is a robust and effective novel GNN for predicting molecular properties. The integration of CNN-GNN in drug discovery presents a powerful and innovative approach to accurately predict and analyze molecular properties. This hybrid model leverages the strengths of both CNNs in handling drug discovery data and GNNs in processing graph-based molecular structures, resulting in a comprehensive feature extraction process that enhances prediction accuracy. The experimental results show the integrated CNN-GNN achieves 97.7%

accuracy, 96.28% precision, 96.2% recall, and 0.938 MCC value which is greater than RNN, traditional CNN, and traditional GNN. By combining CNN and GNN, this novel DL algorithm addresses the complex nature of molecular data, enabling more precise and reliable predictions.

In future, multi-omics data will be used to shed light on the interaction of a drug and its biological processes. This might require more advanced hybrid architectures, real-time drug discovery platforms, or other much advanced methods that can accelerate the identification of potential candidates for a drug. Additionally, advancements in explainable AI, transfer learning, and domain adaptation can improve model transparency and generalizability across diverse datasets.

References

Adham M. Alkhadrawi, Ying Wang, Chun Li, In-silico screening of potential target transporters for glycyrrhetinic acid (GA) via deep learning prediction of drug-target interactions, *Biochemical Engineering Journal*, Volume 181, 2022, https://doi.org/10.1016/j.bej.2022.108375.

Ahmed KT, Park S, Jiang Q et al (2020) Network-based drug sensitivity prediction. *BMC Medical Genomics* 13:193.

Arshed MA, Mumtaz S, Riaz O, Sharif W, Abdullah S (2022) A deep learning framework for multi-drug side effects prediction with drug chemical substructure. *International Journal of Innovative Science and Technology* 4(1):19–31.

Askr, H., Elgeldawi, E., Aboul Ella, H. et al. Deep learning in drug discovery: an integrative review and future challenges. *Artificial Intelligence Review* 56, 5975–6037 (2023). https://doi.org/10.1007/s10462-022-10306-1.

Attwood MM, Fabbro D, Sokolov AV et al (2021) Trends in kinase drug discovery: targets, indications, and inhibitor design. *Nature Reviews Drug Discovery* 20(11):839–861.

Azad AKM, Dinarvand M, Nematollahi A, Swift J, Lutze-Mann L, Vafaee F (2021) A comprehensive integrated drug similarity resource for in-silico drug repositioning and beyond. *Briefings in Bioinformatics* 22(3): bbaa126. https://doi.org/10.1093/bib/bbaa126.

Baptista D, Ferreira PG, Rocha M (2021) Deep learning for drug response prediction in cancer. *Briefings in Bioinformatics* 22:360–379.

Benedek R, Stephen B, Andriy N, Michael U, Sebastian N, Eliseo P (2021) A unified view of relational deep learning for drug pair scoring. *CoRR*. https://arxiv.org/abs/2111.02916.

Betsabeh T, Mansoor ZJ (2021) Using drug–drug and protein-protein similarities as feature vectors for drug–target binding prediction. *Chemometrics and Intelligent Laboratory Systems* 217:104405. https://doi.org/10.1016/j.chemolab.2021.104405.

Bongini P, Pancino N, Dimitri GM, Bianchini M, Scarselli F, Lio P (2022) Modular multi-source prediction of drug side effects with DruGNN. *CoRR*. http://arxiv.org/abs/2202.08147.

Chen, Z. (2020). Applications of artificial intelligence in drug development using real-world data. *Drug Discovery Today*. Advance online publication. doi:10.1016/j.drudis.2020.12.013 PMID:33358699.

Fatehifar M, Karshenas H (2021) Drug–drug interaction extraction using a position and similarity fusion-based attention mechanism. *Journal of Biomedical Informatics* 115:103707. https://doi.org/10.1016/j.jbi.2021.103707.

Feng YH, Zhang SW, Shi JY (2020) DPDDI: a deep predictor for drug–drug interactions. *BMC Bioinformatics* 21:419. https://doi.org/10.1186/s12859-020-03724-x.

Han X, Xie R, Li X, Li J (2022) SmileGNN: drug–drug interaction prediction based on the SMILES and graph neural network. *Life (Basel)*. 12(2):319. https://doi.org/10.3390/life12020319.

He C, Liu Y, Li H, Zhang H, Mao Y, Qin X, Liu L, Zhang X (2022) Multi-type feature fusion based on graph neural network for drug–drug interaction prediction. *BMC Bioinformatics* 23(1):1–8.

Hu J, Gao J, Fang X, Liu Z, Wang F, Huang W, Wu H, Zhao G (2022) DTSyn: a dual-transformer-based neural network to predict synergistic drug combinations. *bioRxiv*. https://doi.org/10.1101/2022.03.29.486200.

Jimenez-Luna, J. (2020). Drug discovery with explainable artificial intelligence. *Nature Machine Intelligence*. doi:10.103842256-020-00236-4.

Kim J, Park S, Min D, Kim W (2021) A comprehensive survey of recent drug discovery using deep learning. *International Journal of Molecular Sciences* 22:9983. https://doi.org/10.3390/ijms22189983.

Klambauer, Günter, Sepp Hochreiter, and Matthias Rarey (2019). Machine learning in drug discovery. *Journal of Chemical Information and Modeling* 59(3), pp. 945–946. doi:10.1021/acs.jcim.9b00136.

KV, Greeshma & Gripsy, Viji. (2021). A review on classification and retrieval of biomedical images using artificial intelligence. *Lecture Notes in Computer Science*. 10.1007/978-3-030-75220-0_3.

KV, Greeshma & Gripsy, Viji. (2020). Image classification using HOG and LBP feature descriptors with SVM and CNN.

Lian M, Du W, Wang X, Yao Q (2021) Drug–target interaction prediction based on multi-similarity fusion and sparse dual-graph regularized matrix factorization. *IEEE Access* 9:99718–99730. https://doi.org/10.1109/ACCESS.2021.3096830.

Liu S, Zhang Y, Cui Y, Qiu Y, Deng Y, Zhang W, Zhang Z (2021) Enhancing drug–drug interaction prediction using deep attention neural networks. *bioRxiv*. https://doi.org/10.1101/2021.03.16.435553.

Luo D, Cheng W, Xu D, Yu W, Zong B, Chen H, Zhang X (2020) Parameterized explainer for graph neural network. *Advances in Neural Information Processing Systems* 33:19620–19631.

Lyu T, Gao J, Tian L, Li Z, Zhang P, Zhang J (2021) MDNN: a multimodal deep neural network for predicting drug–drug interaction events. *Proceedings of the Thirtieth*

International Joint Conference on Artificial Intelligence (IJCAI-21), pp 3536–3542. https://doi.org/10.24963/ijcai.2021/487.

Martin Drance, A. (2021). Neuro-symbolic XAI for computational drug repurposing. *Proceedings of the International Joint Conference on Knowledge Discovery, Knowledge Engineering and Knowledge Management (IC3K 2021)*, 2, 220-225. doi: 10.5220/0010714100003064.

Nag S, Baidya ATK, Mandal A et al (2022) Deep learning tools for advancing drug discovery and development. *3 Biotech* 12:110. https://doi.org/10.1007/s13205-022-03165-8.

Nguyen T, Nguyen TT, Nguyen T, Le DH (2021) Graph convolutional networks for drug response prediction. *IEEE/ACM Transactions on Computational Biology and Bioinformatics* 19:146–154.

Pouryahya M, Oh JH, Mathews JC et al (2022) Pan-cancer prediction of cell-line drug sensitivity using network-based methods. *International Journal of Molecular Sciences* 23:1074. https://doi.org/10.3390/ijms23031074.

Qiu K, Lee J, Kim H, Yoon S, Kang K (2021) Machine learning based anti-cancer drug response prediction and search for predictor genes using cancer cell line gene expression. *Genomics Inform*. https://doi.org/10.5808/gi.20076.

Ren S, Tao Y, Yu K et al (2022) De novo prediction of Cell-Drug sensitivities using deep learning-based graph regularized matrix factorization. *Pacif Symp Biocomput*. https://doi.org/10.7490/f1000research.1118807.1.

Song T, Zhang X, Ding M, Rodriguez-Paton A, Wang S, Wang G. DeepFusion: A deep learning based multi-scale feature fusion method for predicting drug-target interactions. *Methods*. 2022 Aug;204:269-277. doi: 10.1016/j.ymeth.2022.02.007. Epub 2022 Feb 24. PMID: 35219861.

Wang YB, You ZH, Yang S et al (2020) A deep learning-based method for drug–target interaction prediction based on long short-term memory neural network. *BMC Med Inform Decis Mak* 20:49. https://doi.org/10.1186/s12911-020-1052-0.

Yan C, Duan G, Zhang Y, Wu F-X, Pan Y, Wang J (2022) Predicting drug–drug interactions based on integrated similarity and semi-supervised learning. *IEEE/ACM Trans Comput Biol Bioinf* 19(1):168–179. https://doi.org/10.1109/TCBB.2020.2988018.

Yifan D, Xinran X, Yang Q, Jingbo X, Wen Z, Shichao L (2020) A multimodal deep learning framework for predicting drug–drug interaction events. *Bioinformatics* 36:4316–4322

Zhang C, Lu Y, Zang T (2022) CNN-DDI: a learning-based method for predicting drug–drug interactions using convolution neural networks. *BMC Bioinf* 23:88. https://doi.org/10.1186/s12859-022-04612-2 0.

Zhang Y, Ye T, Xi H, Juhas M and Li J (2021) Deep Learning Driven Drug Discovery: Tackling Severe Acute Respiratory Syndrome Coronavirus 2. *Front. Microbiol.* 12:739684. doi: 10.3389/fmicb.2021.739684.

Chapter 9

Comprehensive Data Preparation and Preprocessing Strategies for Effective Drug Discovery

S. Pooja
Vidya Niranjan*
and S. Likitha
Department of Biotechnology, RV College of Engineering, Bangalore, affiliated to Visvesvaraya Technological University, Belagavi, India

Abstract

Drug discovery is the most advanced *in-silico* technology. Initial methods involve data preparation and preprocessing, which are the most crucial steps to ensure the reliability and effectiveness of subsequent analyses. Data collection and source describe the functional, biological, and molecular data set, including high-throughput screening. Data cleaning and quality control are essential to irradicate errors and inconsistencies arising from raw data. Data cleaning and quality control involve rectifying missing values and duplicates to ensure integrity and accuracy. Feature engineering for molecular data has predictive machine learning models. This technique involves chemical descriptors and fingerprints; it helps to examine the biological and chemical properties essential for predictive modeling. Finally, handling imbalanced data in drug discovery, where the number of active compounds exceeds the number of inactive compounds. Resampling methods are cost-effective, and advanced algorithms are available to handle imbalanced datasets

* Corresponding Author's Email: vidya.n@rvce.edu.in

In: Harnessing AI for Drug Discovery
Editor: Kolla Bhanu Prakash
ISBN: 979-8-89530-540-9

effectively. It is essential for developing models that can accurately predict active compounds despite the imbalance. The chapter highlights the significance of integrating multiple data sources to construct an enhanced dataset to increase the robustness of drug discovery for model development.

Keywords: drug discovery, data collection, data cleaning/quality, handling imbalanced data, machine learning

Introduction

Data preparation and preprocessing are the first steps in the drug development process. These stages verify the quality of the data collected for analysis. Data is acquired or created virtually daily at various phases of the drug development process. While collecting data, most of the time, the data obtained contains errors, repetitions, missing numbers, and inaccuracies to avoid blunders throughout the drug discovery process. The data preparation and preprocessing procedure guarantees no duplicates in the data and identifies missing or incorrect information [1]. Because of the large volume of data collected, various methods and procedures must be employed to prepare and preprocess the datasets for further analysis.

Data preparation is the process of preparing the data for analysis. The steps involved in data preparation are data collection, integration, cleaning, transformation, reduction, and splitting. Through data preparation, we can ensure that the dataset's quality and integrity are increased. Good quality data without noise, redundancy, consistency, etc, is essential in developing models using machine learning because the efficacy and precision of ML models are directly affected by the data quality. Data preprocessing techniques are employed to improve the data quality in the datasets. The most typical data preprocessing steps are data integration, enhancements and enrichment, transformation, reduction, discretization, and cleansing [1]. These activities include handling missing values, encoding categorical variables (converting categorical data into numerical format), normalization and scaling, outlier detection and treatment, feature engineering, dimensionality reduction, and data augmentation [2].

Role of Machine Learning in Data Preparation and Preprocessing

In the modern era of drug discovery, Machine learning is successfully employed in various steps of the drug discovery process. Massive datasets and techniques are used to train the ML models, which enable them to learn how to do the task. This training improves the ability to find hidden patterns and correlations in the dataset. ML employs algorithms to read data, learn from it, and then determine or predict the future status of new data sets. ML algorithms are used to automate and optimize the data in the data preparation and preprocessing steps. Since the datasets obtained for data preparation and preprocessing are huge, using algorithms makes detecting and correcting the errors, anomalies, and outliers in the dataset easy. ML algorithms are also widely used in handling missing data to maintain the integrity of the dataset without losing valuable information [3].

Data Collection and Sources in Drug Discovery

A collection of discrete and continuous values that are used to describe the quantity, quality, statistics, and other fundamental units is known as data. In the drug discovery process, a dataset is called large-scale information collected and analyzed throughout the process. This process involves gathering various types of data to support the identification, validation, and development of new pharmaceutical compounds [4].

Source of Data

Drug development in the past decades has generated tons of data about the drug and its activities. Each stage of drug development contributes to various types of data. These datasets are essential for future use in identifying trends and behavior of the drug at multiple stages.

The data collected from the experimental and theoretical processes differ in various stages of drug development. These datasets are unique and relevant to the drug's development stage. The source of a dataset defines the place where the datasets are available. There are many open-source databases available for the public to access the data. Databases typically contain information on a particular stage of drug discovery or specific data relevant to drug development. Table 1 lists the databases and sources that are relevant to the development and discovery of new drugs.

Table 1. The databases and their respective sources

Databases	Sources
Genomic and Proteomic Databases	GenBank, Ensembl, UniProt
Chemical and Pharmacological Databases	PubChem, ChEMBL, DrugBank
High-Throughput Screening (HTS) Platforms	National Center for Advancing Translational Sciences (NCATS)
Clinical Trial Databases	ClinicalTrails.gov, European Clinical Trials Database (EudraCT)
Electronic Health Records (EHRs)	HealthCare Databases-IBM Watson Health, Optum EHRs
Patient Registries and Biobanks	National Institutes of Health (NIH) Biobank
Omics Data (Genomics, Transcriptomics, Proteomics, Metabolomics)	The Cancer Genome Atlas (TCGA)
Collaborative Networks and Consortia	The Innovative Medicines Initiative (IMI), The Structural Genomics Consortium (SGC)
Real-World Evidence (RWE) Sources	Patient-reported outcomes, Health Insurance Claims Data
Public Health and Epidemiological Data	World Health Organization (WHO), Centers for Disease Control and Prevention (CDC)
Computational and Bioinformatics Tools Data	Molecular dynamic simulations-molecular interactions data, Docking studies data
Literature and Patents	PubMed, Google Patents, Google Scholar

Data Source

In research, various methods are used to collect data, but all these methods fall into two categories: primary data and secondary data.

Primary data is original data obtained for a particular study aim, while secondary data is just the analysis and interpretation of the primary data [5].

Primary Data Source

Primary data are original sources that include clinical trial data (i.e., Phase I, Phase II, Phase III, & Phase IV), laboratory research data (i.e., *in-vitro, in-vivo* studies), Surveys, Questionnaires, and Patient registries.

- *Survey and Questionnaires*: In drug development, surveys and questionnaires collect data. A "survey" collects data from a designated group of respondents to gather information and insights about various topics of interest. A “questionnaire,” a set of specific questions, is used to gather information from respondents. The questions can be Open-ended or Close-ended. The questionnaire also contains “Demographic questions” to collect information on the respondents and “Content questions” to focus on the main topic of interest. The data collected from surveys and questionnaires include information on the adverse effects of the drug, the treatment efficacy of the drug, understanding patient adherence to medication regimens, and more.

- *Clinical trials*: Clinical trial data is considered primary data since each phase of the clinical trial produces relevant data that is used for future research purposes. Phase I of the clinical study collects data on the test’s safety, dose, and side effects in a small group of healthy participants/volunteers. Phase II and Phase III involve data collection from larger groups of unhealthy people on the efficacy and the side effects of the drug on them, along with comparing the drug with commonly used treatments. In Phase IV, data is gathered on the marketed drug to determine its impact on diverse demographics and any adverse effects associated with long-term usage.
- *Laboratory research*: The most essential experimental methods for investigating biological processes, drug effects, and disease mechanisms are through *in-vitro* and *in-vivo* studies. In-vitro studies are conducted to understand the mechanism of the drug, which leads scientists to identify potential drug candidates. Once potential candidates are identified, toxicity levels are tested to shortlist the drugs with more potential. In-vivo studies are performed to evaluate the drug further. In in-vivo studies, the efficacy and safety of the drug in a complex biological system is studied. Further, the pharmacokinetics and pharmacodynamics of the drug and its potential effects on the biological body are also studied. Hence, the data collected from laboratory research is considered primary and used in further drug development.

Secondary Data Source

As mentioned, secondary data is the information that a researcher gathers to conduct new research from information already published by someone else. Secondary data is more refined than primary data. Secondary data collection is more accessible and rapid as the data is already collected and processed by someone else [5]. Secondary data sources include medical and health records, published literature, databases, repositories, biobanks, genomic databases, market research reports on drugs, and more.

Stages of Data Collection

The study and development of new pharmaceuticals have resulted in enormous amounts of diverse data that are constantly increasing and are gathered from several distributed sources and tools. This information comes from several stages of the drug discovery process, where data is only collected to develop drugs targeted to particular illnesses. Each domain has differences in the quantity and quality of data [6].

Target Identification and Validation

The first stage in drug discovery is determining a disease's biological origin and possible therapeutic targets. Potential targets in target identification include genes, nucleic acids, and proteins. The first stage in target identification is establishing the target's function and role in the disease. Next, the target validation ensures the target has a potential therapeutic benefit. This step is considered necessary in drug development for the success of a drug [7].

A specific type of data is required to identify and validate the target. Genomic, proteomic, and nucleic acid data are the three essential categories of necessary data. Each piece of data is vital in the drug development process. Genomic data provides information on genes, variants, and their location within the genome of a particular organism. Genomic data is used to discover genes linked to diseases. Proteomic data consists of a set of proteins produced by a biological system. Proteomic data is used to identify proteins and their roles in particular diseases. Nucleic acid data consists of information related to sequences and structures of nucleic acids. The nucleic acid data is used to validate the role of a gene or genetic pathway as a therapeutic target.

Lead Compound Identification

In drug development, the lead compound identification step identifies chemical compounds with potential lead compounds. The data required for lead compound identification is obtained from high-throughput screening (HTS) of large libraries of compounds with biological activity. Along with HTS, virtual screening, and molecular docking simulations, ML and deep learning (DL) also aid in identifying potential drug candidates. Chemical libraries are another source of data for lead compound identification. It includes information on a range of chemical compounds that are utilized to determine the potential of the chemical compound to be effective drug candidates. Once a set of potential lead compounds is identified, *in-silico* techniques predict drug candidate interactions and target interactions.

Lead Optimization

Lead optimization is the process of developing a therapeutic candidate after identifying an initial lead molecule. The approach involves an iterative sequential synthesis and characterization of a potential medication to offer an overview of the relationship between chemical structure and activity regarding target interactions and metabolism [7].

In the process of lead optimization, pharmacokinetic and pharmacodynamic (PK/PD) data and Structure-Activity Relationship (SAR) Data are obtained. SAR data explains the relationship between a compound's molecular structure and biological activity. In contrast, PK/PD data is analyzed to understand more about the function of the drugs on the body.

- *Preclinical testing:* during the preclinical testing process, the potential lead compound is tested for its efficacy and toxicity. The testing is done in *vitro* and *in vivo*, and toxicological data is used using cell culture, animal studies, and other experimental methods. Once the testing is done, it produces data relevant to in-vivo and in-vitro studies. This data is vital in finding the most promising drug candidate for the following research step.
- *Clinical trials:* the safety, effectiveness, and dose of the lead compound are calculated in clinical trials. Clinical trials produce a massive amount of data in different phases. The data mainly includes demographics and study-specific measures. Data is generated in raw form during clinical trials, which is later abstracted, coded, and transcribed. The obtained data is cleaned into an analyzable data set. The scientific communities and public are given access to the datasets, which include a small subset of the data collected and analyzed during the trial [8].
- *Regulatory data:* once the data is collected from various steps of the drug discovery process, the regulatory data ensures the drug can be produced with known quality, purity, and strength.

- *Post-market surveillance:* A significant amount of data is collected to monitor a drug's performance in the real world after its approval. One such example is real-world evidence (RWE). The RWE data helps identify the drug's adverse effects and safety issues that may not have been found during the clinical trials. Hence, analyzing RWE is essential to assess the drug's risk-benefit profile and ensure patients' safety.

Role of ML in Data Collection in Drug Discovery Process

The drug discovery process produces large datasets. These data include statistics, tables, abstract ideas, concrete measurements, analysis, images, calculation, or a discussion. The origin of these data comes from the drug research and development phase, especially during clinical trials. Advanced machine learning algorithms and tools have been developed and used to make analyzing this vast amount of data more manageable. Researchers and pharmaceutical companies rely on ML and DL algorithms and models to increase the accuracy of the data in less time.

Supervised and unsupervised learning approaches are the two categories of machine learning methodologies. Unsupervised approaches are used to develop exploratory models, allowing for data clustering in ways the user does not define. Meanwhile, supervised learning methods are used to create training models that predict future values of data categories or continuous variables.

Researchers used machine learning (ML) to examine massive datasets, including information on the function of a potential target, to build predictions about causality [3]. Several research projects employed ML methods in the drug discovery process. This approach is used to mine literature papers, identify novel targets, improve therapeutic target identification success rates, analyze large data sets of small molecules to significantly reduce data, discover biomarkers, and develop drug sensitivity models using preclinical data sets. Using ML can boost the speed, accuracy, and efficiency of data collection in the drug discovery process when combined with methods like convolutional neural network (CNN), deep learning (DL), recurrent neural network (RNN), deep autoencoder neural network (DAEN), natural language processing (NLP), support vector machine (SVM), deep neural network (DNN), and many other models and algorithms [9–13].

Data Storage and Management

Data storage in the drug discovery process includes the data obtained from all the steps and stages of drug discovery. Each stage generates data that are complex and huge, and the storage criteria differ from each other. In the pharmaceutical industry, SAP EWM (Extended Warehouse Management) stores vast amounts of data generated during drug development [14]. SAP EWM is considered a more modern and flexible warehouse that can run in the cloud; hence, managing high-volume data is more effortless.

Database

A database is a structured data set in the computer and managed by a software Database Management System (DBMS). DBMS provides consistency, safety, and integrity to the data in the database [15]. There are four types of models for databases: Relational DBMS, NoSQL DBMS, Object-oriented DBMS, Hierarchical DBMS, and Network DBMS. However, using a database certainly has drawbacks, as data managed and stored within the transactional system could be more user-friendly. A data warehouse was developed to overcome this problem. Data-warehouse allows users to access the data without interfering with day-to-day system operation [16, 17].

Data Warehouse

A data warehouse is a computer file and an analytical database used as a decision-support system. A data warehouse is a collection of data that is subject-oriented, integrated, time-variant, and non-volatile. It is used to assist management decisions [18]. A data warehouse collects and stores subject-oriented, integrated, time-variant, and non-volatile data to aid management decision-making [16]. During the analytical process, this type of data warehouse is not used for transaction processes but is intended for application. The data warehouse is designed for query and analysis, not transaction processing. The data warehouse stores data efficiently so that users can access it. Data warehouses contain large volumes of data to be handled by databases, which offer a flexible way of dealing with and retrieving the data. Relational and Multidimensional databases are two categories of databases that contain and manage data in the Data warehouse [19].

The Key Characteristics of a Data Warehouse

- *Subject-oriented:* The data warehouse is organized around critical subjects like patients, clinical trial data, and drug efficacy. It does not manage data related to business functions.

- *Integrated:* Multiple transactional systems and external sources merge their data into the data warehouse. This ensures consistency and coherence of the data.
- *Time-variant*: Data warehouse maintains historical data to track changes over time, which is crucial for trend analysis.
- *Non-volatile*: Data is unmodifiable; users get only read-only access, preserving the historical integrity of information.

A few of the best examples of Data warehouses are InterMine, TargetMine, DW4TR, The Cancer Genome Atlas (TCGA), and IMI's eTRIKS data warehouse. Intermine is an open-source data warehouse system that integrates and analyzes various biological data types [20]. TargetMine, utilizing the InterMine framework, focuses on drug discovery, integrating relevant datasets to facilitate target identification and prioritization [21]. The TCGA collaborates with the National Cancer Institute (NCI) and the National Human Genome Research Institute (NHGRI) to catalog cancer-causing genetic mutations using a large-scale data warehouse. IMI's eTRIKS (European Translational Information and Knowledge Administration Services) is an Innovative Medicines Initiative (IMI)-funded initiative that provides data administration and analysis help for translational research projects across Europe. A lightweight, patient-centric clinical data model and a specimen-centric genetic data model are used to build the Data Warehouse for Translational Research (DW4TR) [22].

Data Cleaning and Quality Control

Importance of Data Cleaning in Drug Discovery

Researchers clean raw data for analysis during the drug discovery process instead of using it directly for further study. Raw data consists of errors, inconsistencies, and inaccuracies removed through data cleaning. The cleaned data is more reliable and gives valid results in the drug discovery process. Clean data enhances the reliability of statistical analyses and predictive models. Clean data also minimizes the biases that could affect the interpretation of the results [23]. Most traditional data cleaning methods rely on user input to identify irregularities and remove erroneous values.

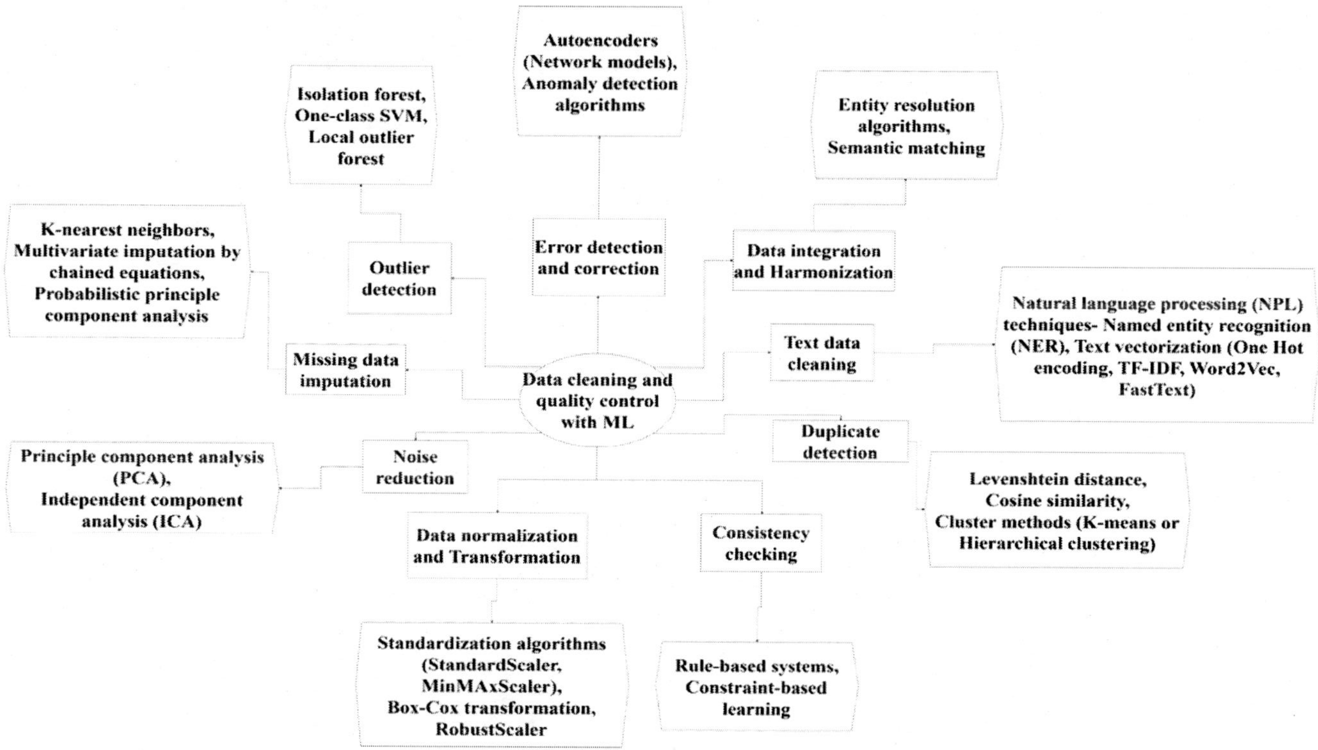

Figure 1. Data-Cleaning Process Involved in ML Algorithm.

Several tools and techniques are available to prepare data for analysis. A few such tools are Microsoft Excel or Google Sheets, which clean primary data and remove raw data. Data errors can be detected and repaired using statistical tools such as R, SAS, and SPSS. Data cleansing tasks can be automated using Python and R programs. OpenRefine, Trifacta, and Talend are utilized for data cleaning and transformation, followed by ML algorithms for advanced and complex data challenges. Figure 1 represents the data-cleaning process.

Data Cleaning for Machine Learning

A user's qualitative assessments of data quality often depend on the outputs at the application level. Data must be cleaned before ML is used for data cleaning and quality control, as low-quality data can result in poor-quality models. ML models are susceptible to mistakes at the population level, which might bias the model [24].

Managing data quality in machine learning systems requires the capacity to make well-informed cleaning decisions at various stages of the program, for multiple types of data and mistakes, and for different objectives as the models, workflow, and data change. However, the data cleaning process must be repeated through ML numerous times to attain an equilibrium point when the original data is sufficiently cleaned to represent the actual distribution, free of biases and errors.

Data cleaning and quality control typically consist of two stages: error identification and error repair. Machine learning techniques assist in identifying and correcting mistakes, handling missing data, and guaranteeing consistency across big and complicated datasets [23].

1. *Outlier detection/novelty detection/anomaly detection:* An outlier is an unusually high or low-value data point that differs considerably from the other observations in a dataset. There are two kinds of outliers. Univariate and multivariate outliers can be caused by data collecting or recording mistakes and differences between variables collected from biological or chemical experiments. ML algorithms and techniques such as Isolation Forest, One-class SVM, and DBSCAN (Density-Based Spatial Clustering of Applications with Noise) can help identify and correct these errors. Isolation forest (iForest) finds outliers effectively by separating observations using

random forests; however, iForest is only sensitive to global outliers and is ineffective at dealing with local ones [25]. A tree structure is created by subsampling data by applying random cuts to the values of randomly chosen features from the dataset to process the data in an Isolation Forest (iForest). Samples that reach deeper into the tree branches are less likely to be abnormal, while shorter branches suggest an abnormality [26].

2. One-class SVMs are an extension of SVMs for unlabelled data that is used to detect anomalies. The one-class SVM identifies data points that differ considerably from the norm [27]. Using a non-linear mapping function, one-class SVM transforms the sample points into a higher-dimensional feature space and then constructs a hyperplane that separates the resulting image points from the origin by the most significant margin. Here, the origin is regarded as the only member of the other class [28]. DBSCAN (Density-Based Spatial Clustering of Applications with Noise) is an unsupervised learning approach that involves cluster analysis. It finds outliers using density clustering. DBSCAN is sensitive to its input parameters, which substantially impact the clustering outcome; hence, an automated and more exact approach for determining the input parameters is required [29]. Local outliner forest is another method for detecting density-based outliers. LOF does not assume any underlying or predefined distribution to handle dense data. It also discovers the dataset in different densities. A Specific drawback of LOF is that a significant amount of memory is required to compute the distance between points, which impacts execution time. Also, it needs help to handle the succession of outliers. Many researchers have created novel approaches to address the shortcomings of the IF and LOF methodologies [26, 30].
3. *Missing data imputation:* Missing data is a standard error during clinical trials. Missing data can occur in any research design, including randomized controlled trials, cohort studies, and case-control studies. Missing values are unavoidable, especially in studies involving frequent data gathering. Three general categories can be used to describe missing data: missing not at random (MNAR), missing at random (MAR), and missing completely at random (MCAR). There are many approaches to avoiding missing data; one such approach to managing missing data is simply excluding missing data from further analysis [9]. However, eliminating data variables with missing values is ineffective for dealing with missing data [31].

Many studies utilize the K-nearest neighbors (k-NN) method to fill in missing values by averaging the values of the k-nearest neighbors [32]. Multivariate imputation by chained equations (MICE) is a regression model that uses a multiple imputation (MI) approach to handle missing data by progressively substituting it over numerous repetitions. MI generates an estimate by performing statistical analysis on available information and dealing with the uncertainty created by the existence of missing data [33]. Another machine learning approach, Probabilistic Principal Component Analysis (PPCA), presumes missing values when the original dataset is recovered from its dimension-reduced version. PPCA uses the Expectation-Maximization (EM) technique to repeatedly construct the Maximum Likelihood Estimates (MLE) of an incomplete dataset [34].

4. *Error detection and correction:* Errors are expected in the data. Errors in input data lead to errors in the output. ML techniques are used to minimize the data errors. Another big problem faced while developing an ML model is that giving incorrect data as input to an ML model reduces its performance. Hence, anomaly detection algorithms are utilized to discover and fix errors. Anomaly algorithms identify and correct erroneous data entries by learning from the data distribution. Neural network models are another approach to encoding data into a lower-dimensional space and then reconstructing it, detecting deviations along the way.
5. *Duplicate detection:* Another type of data that hinders the output is duplicates. Manual elimination of duplicates can be done using Google Spreadsheets or Microsoft Excel. However, Cluster methods such as K-means or hierarchical clustering are highly favorable when dealing large amounts of data. Cluster methods may group comparable data/records and discover duplicates in datasets. Similarly to cluster methods, Levenshtein distance and cosine similarity are fuzzy matching approaches used to detect and combine duplicates in datasets.
6. *Noise Reduction:* data noise stems from low-level data errors resulting in inaccurate or irrelevant information. Hence, noise reduction is significant in the data-cleaning process. While data preprocessing includes cleaning, normalizing, and removing outliers, the data quality is enhanced, and noise is reduced to the greatest extent. Traditional outlier detection methods used in noise reduction

are local outlier factors, distance-based, and clustering-based [35, 36]. ML algorithms such as Principal component analysis (PCA), Independent component analysis (ICA), Support vector systems, Regression, and Bayesian techniques are also used in reducing noise in the data [37]. While noise gives inaccurate results and confuses important patterns, noise in data can also affect the model quality developed using ML. Some types of noise that are present during ML modeling are feature noise, systemic noise, random noise, and background noise. Noise elimination is necessary to improve ML models' robustness and general performance. Specific ways to handle noises include Fourier transform, Autoencoders, PCA and ICA methods, and constructive learning methods [38].

Feature Engineering for Molecular Data

When creating a predictive model using statistical modeling or machine learning, feature engineering is essential for identifying and transforming the most significant variables from the raw data. Feature engineering comprises five processes: feature creation, transformations, feature extraction, exploratory data analysis, and benchmarking [39–42].

"Feature engineering for molecular data" refers to converting raw molecular data into meaningful features or variables that may be utilized successfully in machine learning models. One of the feature engineering processes called "transformations" involves altering the predictor variables to enhance the model's functionality. Some of the benefits include making the model more adaptable to the types of data it can process, ensuring that variables are on the same scale, which makes the model easier to understand, increasing accuracy, and preventing computational errors by guaranteeing that all features fit within the model's acceptable range [39, 43].

In the drug discovery process, feature engineering involves extracting and constructing relevant attributes from molecular structures, chemical properties, or biological data to improve the performance and prediction of developed computational models. Principal Component Analysis (PCA) or t-SNE (t-Distributed Stochastic Neighbour Embedding) are two feature engineering techniques that help reduce the dimensionality of high-dimensional molecular data while retaining essential data.

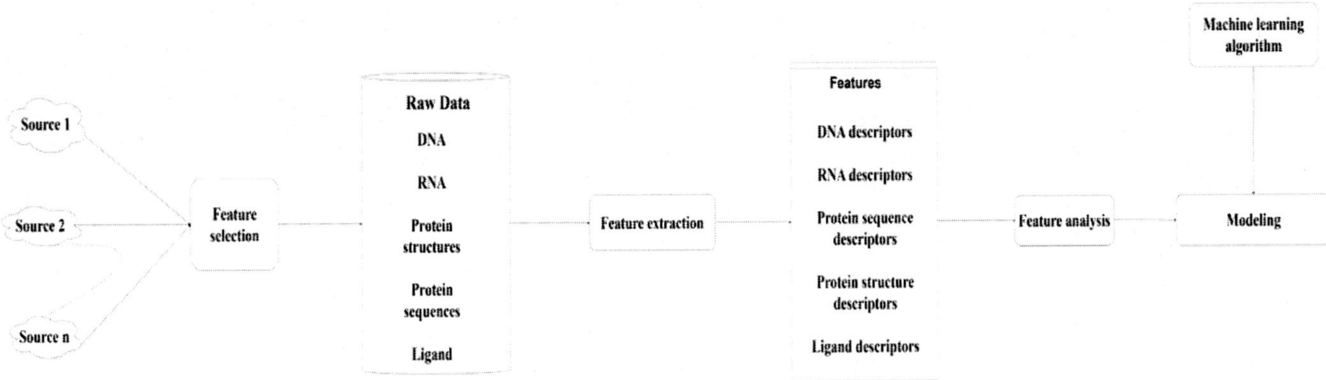

Figure 2. Feature Engineering Model.

Feature engineering also uses min-max scaling, Z-score normalization, and logarithmic transformations to create standard and normalized features from molecular data, ensuring consistency across different datasets and experiments [39, 44, 45]. Engineering features from molecular data can aid in identifying potential drug targets by highlighting important chemical or biological properties that correlate with target interactions. During the lead optimization stage, feature engineering enhances and improves lead compounds by optimizing effectiveness, safety, and pharmacokinetics characteristics. Another benefit of feature engineering is developing prediction models that evaluate the attributes of ADMET (Absorption, Distribution, Metabolism, Excretion, and Toxicity) based on molecular features. These models aid in selecting and improving drug candidates [46, 47]. Figure 2 represents the feature engineering model.

- *Feature Creation:* It is an efficient method of creating new attributes that capture more critical information in a dataset than the original attributes. Three methods used in feature creation are domain-specific feature extraction, data mapping to a new space using Fourier transformation, wavelet transformation, etc., and feature construction, which combines existing features. Feature creation improves the model's performance, robustness, interpretability, and flexibility [48, 49].
- *Feature Transformation:* It is a process that involves transforming the features into a more appropriate representation. Types of feature transformation are normalization, scaling, encoding, and transformation. The scaling technique transforms the numerical variables to have a similar scale, making comparing them more accessible. While encoding transforms categorical features into numerical representation, transformation helps transform features using mathematical operations to modify the distribution or magnitude of the features [50, 51].
- *Normalization and standardization:* Normalization impacts a dataset's values of numeric columns. It adjusts them to a standard scale without changing differences in the value range [52]. This process ensures that no single feature overpowers others due to scale differences. It also provides all the features that contribute to the model. Normalization is used to preprocess data when the data distribution is unknown or non-Gaussian [53]. Normalization scales

the values of a feature to a specific range, often between 0 and 1. In contrast, standardization is used when the data is unknown or assumed to be Gaussian. Standardization, or z-score normalization, adjusts the features to have a mean of zero and a standard deviation of one [54].

Importance of Feature Engineering in Drug Discovery

- *Improving model performance*: feature engineering enhances the accuracy and robustness of predictive models. Feature engineering helps create more relevant and informative features, crucial for identifying potential drug candidates and predicting their efficacy and safety [55].
- *Handling Complex Biological Data and Enhancing Interpretability:* In drug discovery, feature engineering involves collecting and constructing relevant attributes from molecular structures, chemical properties, or biological data. This data is then used to improve the performance of the computational models [56]. Features developed from feature engineering provide insights into the underlying biological mechanisms and relationships. Through this, researchers understand why a particular compound is effective or specific biological responses occur, etc, aiding in hypothesis generation and validation.
- *Data integration and Optimization*: data from the drug discovery process comes from many sources. These diverse sources are genomic databases, chemical libraries, clinical trials, and electronic health records. Feature engineering plays a significant role in enabling the integration and harmonization of these diverse datasets, allowing for complete analysis. By choosing and developing the most valuable features, this process cuts down the computational burden and enhances the efficiency of machine learning models. This process is especially significant given the vast data on drug development.

Handling Imbalanced Data in Drug Discovery

Handling imbalanced data is a common challenge in drug discovery, as datasets often contain a significantly higher number of negative samples (e.g., inactive compounds) than positive samples (e.g., active compounds). The dataset utilized in drug development is considered to be unbalanced if it belongs to the minority class and seems sparse when compared to the majority class, also known as the positive class. Imbalanced data is a typical challenge in the drug development process, and it happens when certain types of data distributions dominate the instance space more than others. An imbalanced class can be dealt with by handling and solving the imbalanced class problems in machine learning, such as the data sampling approach given in Figure 3, the cost-sensitive learning approach, and the ensemble approach [57, 58].

- *Resampling:* Resampling is a technique for adjusting the class distribution in a dataset where one class (minority class) is notably under-represented relative to another (majority class). Under-sampling and over-sampling are two popular approaches in resampling. In the random under-sampling strategy, the majority class is depopulated at random until it reaches the same size as the minority class. In the Over-sampling approach, the minority class is frequently replicated till a balanced class distribution is achieved. Two approaches to over-sampling are random over-sampling and the synthetic minority oversampling technique (SMOTE) [58].
- *Algorithmic technique:* The oversampling technique addresses class imbalance in datasets by increasing the number of instances in the minority class. Algorithm techniques widely used for oversampling are SMOTE, ADASYN, Borderline-SMOTE, SMOTE-ENN, and SVMSMOTE. Chawla et al. presented SMOTE in 2002 as a method for adding synthetic minority class observations using the k-nearest neighbors algorithm [59].
- *Synthetic data generation- SMOTE (Synthetic Minority Over-sampling Technique):* SMOTE is an over-sampling technique in which a synthetic example is created by interpolating between existing minority class instances [60]. A few well-known SMOTE techniques are Borderline-SMOTE, AHC, ADASYN, Safe-Level-SMOTE, SVMSMOTE, DBSMOTE, ROSE, MWMOTE, and MDO [61]. To improve an ML model's performance, we use SMOTE.

SMOTE enhances the performance of the minority class by reducing bias and balancing the class distribution. The synthetic instances SMOTE creates offer more variability and reduce the risk of overfitting. However, there are certain drawbacks to using SMOTE. SMOTE can introduce synthetic noise by generating samples; it may create samples that fall between classes near the decision boundaries, leading to misclassification [62].

The general steps involved in SMOTE are:

1. Identify minority class instances.
2. Find the k-nearest neighbors for each minority class instance.
3. Generate synthetic samples: Create synthetic samples by randomly selecting one of each minority class instance's k-nearest neighbors. Interpolate the two instances to create a new synthetic instance.
4. Add synthetic samples to the dataset to the minority class to balance the dataset.

ADASYN (Adaptive Synthetic Sampling): Similar to SMOTE, ADASYN creates synthetic observations that follow a straight line between a minority class observation and its k-nearest minority class neighbors [63]. It adjusts the sampling distribution based on the learning difficulty of different regions in the feature space. ADASYN was developed as an extension of SMOTE. Several studies have indicated that SMOTE outperforms ADASYN. ADASYN performs better when the learning model is a Decision tree and has slightly lesser performance variation than SMOTE [64, 65].

The general steps involved in ADASYN are:

1. Calculating class imbalance by determining the ratio of minority class instances to majority class instances.
2. Identifying k-nearest neighbors (often k=5) from both the majority and minority classes for each minority class instance.
3. Compute the density distribution by calculating the ratio of majority-class neighbors to the total number of neighbors.
4. Density distribution is used to determine the number of synthetic samples. Instances with higher difficulty (higher ratio of majority class neighbors) will have more synthetic samples generated.
5. Generating synthetic samples for each minority class instance by interpolating between the instance and its minority class neighbors.

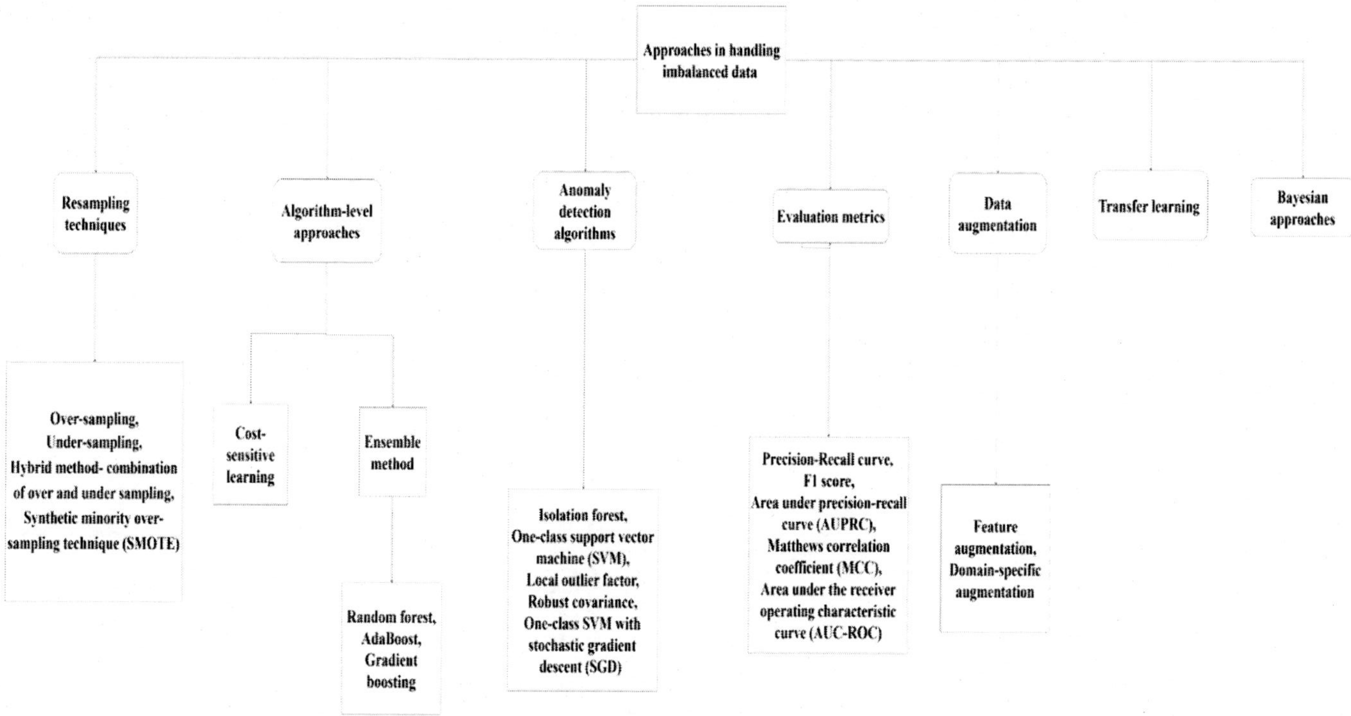

Figure 3. Approaches involved in data handling.

Approaches to Handling Imbalanced Datasets

1. *Data resampling approaches:* to avoid misleading performance of ML models, we employ over-sampling, under-sampling, or both methods in data resampling. In the under-sampling approach, the majority class is selectively depopulated, excluding the sensitive information about the majority class while depopulating.
2. *Ensemble method:* In the Ensemble approach, different classifiers are trained on the imbalanced dataset, and the results are pooled to provide a new result. It's an algorithm-level approach, along with coast-sensitive learning as another approach. Ensemble learning performs better with multiple classifiers than with a single classifier. Multiple classifiers can give a lesser rate of error, which implies more accurate model building. Specific ways of implementing the Ensemble approach are Boosting, AdaBoost, Random Subspaces, Random Forests, and Bagging [58].
3. *Evaluation metrics for imbalanced data*: Evaluation metrics are used to quantify the performance of a predictive model [66]. While evaluating an ML model on imbalanced datasets, traditional accuracy is not sufficient because this accuracy can be misleading when the classes are imbalanced. Hence, evaluation metrics are crucial and suited explicitly for imbalanced data. There are hundreds of evaluation metrics, but researchers have divided evaluation metrics into three groups: Threshold metrics, Ranking metrics, and Probability metrics [67]. Precision, recall, the F1 score, and the AUC-ROC are the most popular evaluation metrics. Threshold metrics evaluate the performance of a classifier based on a specific decision threshold. The most common threshold matrices are accuracy, precision, recall, F-measure/F1-score, Specificity, and Balanced accuracy. Ranking metrics evaluate the performance of a model by considering the ranking of predicted probabilities rather than the binary class labels [68,69]. ROC Curve (Receiver Operating Characteristic Curve), AUC-ROC (Area Under the ROC Curve), Precision-Recall Curve, and AUC-PR (Area Under the Precision-Recall Curve) are a few widely used Ranking metrics [69,70]. The last group, Probability metrics, evaluates the quality of predicted probabilities directly without converting them into class labels or ranks. Examples of Ranking metrics are Logarithmic Loss (Log Loss

or Cross-Entropy Loss), Brier Score, and Root-mean-squared error [71–73].

Other approaches to solving imbalance class problems are:

- Anomaly detection algorithms-isolation forest and One-class SVM
- Data augmentation- feature augmentation and domain-specific augmentation
- Transfer learning
- Bayesian approaches

Conclusion

In conclusion, this paper discusses the data preparation and preprocessing steps and methods in detail. Data collection is the initial step in the process, where the data source can be divided into primary and secondary data sources, including the required data for the drug discovery process. The drug discovery process includes several stages, resulting in large volumes of diverse datasets. ML is more reliable in handling this enormous amount of data, which helps access and organize the massive amount of data. Data warehousing is a popular method that has been used for many years. It represents a centralized repository that stores large volumes of structured and unstructured data from multiple sources. ML techniques also assist in data cleaning and quality control by identifying and correcting mistakes, handling missing data, reducing noise, and guaranteeing consistency across big and complicated datasets. To aid ML models, feature engineering plays a significant role in selecting and transforming raw data while developing ML models. Feature creation and feature transformation in the feature engineering process involve creating new attributes that capture essential information in a dataset rather than the original attributes and transforming the features into a more suitable representation.

Feature engineering minimizes the computational burden and enhances the efficiency of ML models, which is a significant part of the drug development process. Another obstacle while dealing with datasets is imbalanced data that contains significantly more negative samples than positive ones. Resampling methods in ML, such as SMOTE and ADASYN, are widely used to handle imbalanced data. SMOTE and ADASYN are over-sampling techniques in which SMOTE creates a synthetic example by interpolating between existing minority class instances. ADASYN adjusts the sampling distribution based on the learning difficulty of different regions in the feature space. There are several approaches to handling imbalanced data that help avoid the misleading performance of ML models. Hence, data preparation and preprocessing are essential steps that ensure the quality and accuracy of data for analysis and machine learning tasks.

Acknowledgments

The authors would like to thank Akshay Uttarkar for his crucial insights in improving the presentation. Our heartfelt thanks to the staff and administration at R V College of Engineering for their support. Sincere gratitude to Savitribai Jyotirao Phule Fellowship for Single Girl Child, University Grant Commission, for giving fellowship to S. Pooja, PhD candidate.

Author Contribution

VN was involved in the manuscript's ideation, conceptualization, and refining. PS and LS contributed to the overall writing the manuscript. All the authors have read the final version of the manuscript and agreed to the publication.

Funding

No Funding.

Conflict of Interest

The authors declare no conflict of interest.

Ethics Statement

No ethical consent is required.

References

[1] Taleb I, Dssouli R, Serhani MA. Big Data Pre-processing: A Quality Framework. In: *2015 IEEE International Congress on Big Data* [Internet]. New York City, NY, USA: IEEE; 2015 [cited 2024 Jul 12]. p. 191–8. Available from: http://ieeexplore.ieee.org/document/7207219/.

[2] Jamshed H, Khan MSA, Khurram M, Inayatullah S, Athar S. Data Preprocessing: A preliminary step for web data mining. *3C Tecnología* [Internet]. 2019 May 17 [cited 2024 Jul 12];206–21. Available from: https://www.3ciencias.com/articulos/articulo/data-preprocessing-a-preliminary-step-for-web-data-mining.

[3] Vamathevan J, Clark D, Czodrowski P, Dunham I, Ferran E, Lee G, et al. Applications of machine learning in drug discovery and development. *Nat Rev Drug Discov* [Internet]. 2019 Jun [cited 2024 Jul 12];18[6]:463–77. Available from: https://www.nature.com/articles/s41573-019-0024-5.

[4] Chan KCC. Big data analytics for drug discovery. In: *2013 IEEE International Conference on Bioinformatics and Biomedicine* [Internet]. Shanghai, China: IEEE; 2013 [cited 2024 Jul 12]. p. 1–1. Available from: https://ieeexplore.ieee.org/document/6732448/.

[5] Ajayi, V. O. Primary sources of data and secondary sources of data. *Benue State University*. 2017;1[1], 1–6.

[6] Kozyraki R, Kristiansen M, Silahtaroglu A, Hansen C, Jacobsen C, Tommerup N, et al. The Human Intrinsic Factor-Vitamin B12 Receptor,Cubilin: Molecular Characterization and Chromosomal Mapping of the Gene to 10p Within the Autosomal Recessive Megaloblastic Anemia [MGA1] Region. Blood [Internet]. 1998 May 15 [cited 2023 Jul 3];91[10]:3593–600. Available from: https://ashpublications.org/blood/article/91/10/3593/174973/The-Human-Intrinsic-Factor Vitamin-B12.

[7] Deore AB, Dhumane JR, Wagh R, Sonawane R. The Stages of Drug Discovery and Development Process. *Asian J Pharm Res Dev*. 2019 Dec 15;7[6]:62–7.

[8] Lo B. Sharing Clinical Trial Data: Maximizing Benefits, Minimizing Risk. *JAMA* [Internet]. 2015 Feb 24 [cited 2024 Aug 2];313[8]:793. Available from: http://jama.jamanetwork.com/article.aspx?doi=10.1001/jama.2015.292.

[9] Rifaioglu AS, Atas H, Martin MJ, Cetin-Atalay R, Atalay V, Doğan T. Recent applications of deep learning and machine intelligence on in silico drug discovery: methods, tools and databases. *Briefings in Bioinformatics* [Internet]. 2019 Sep 27 [cited 2024 Jul 12];20[5]:1878–912. Available from: https://academic.oup.com/bib/article/20/5/1878/5062947.

[10] Chen H, Engkvist O, Wang Y, Olivecrona M, Blaschke T. The rise of deep learning in drug discovery. *Drug Discovery Today* [Internet]. 2018 Jun [cited 2024 Jul 12];23[6]:1241–50. Available from: https://linkinghub.elsevier.com/retrieve/pii/S1359644617303598.

[11] Jiao Y, Du P. Performance measures in evaluating machine learning based bioinformatics predictors for classifications. *Quant Biol* [Internet]. 2016 Dec [cited 2024 Jul 12];4[4]:320–30. Available from: https://onlinelibrary.wiley.com/doi/10.1007/s40484-016-0081-2.

[12] Mohr DC, Zhang M, Schueller SM. Personal Sensing: Understanding Mental Health Using Ubiquitous Sensors and Machine Learning. *Annu Rev Clin Psychol* [Internet]. 2017 May 8 [cited 2024 Jul 12];13[1]:23–47. Available from: https://www.annualreviews.org/doi/10.1146/annurev-clinpsy-032816-044949.

[13] Esteva A, Robicquet A, Ramsundar B, Kuleshov V, DePristo M, Chou K, et al. A guide to deep learning in healthcare. *Nat Med* [Internet]. 2019 Jan [cited 2024 Jul 12];25[1]:24–9. Available from: https://www.nature.com/articles/s41591-018-0316-z.

[14] Rajora, N. Pharmaceutical drug launch and its readiness in enterprise systems. *Aquat Microb Ecol* [Internet]. 2022;10[5]. Available from: http://www.ijarcsms.com/.

[15] Sharma A, Karamchandani A, Dave D, Patel A, Doshi N. Database Management Systems—An Efficient, Effective, and Augmented Approach for Organizations. In: Senjyu T, Mahalle PN, Perumal T, Joshi A, editors. *ICT with Intelligent Applications* [Internet]. Singapore: Springer Singapore; 2022 [cited 2024 Jul 12]. p. 465–78. [Smart Innovation, Systems and Technologies; vol. 248]. Available from: https://link.springer.com/10.1007/978-981-16-4177-0_47.

[16] Harrington JL. Data Warehousing. In: Relational Database Design [Internet]. *Elsevier*; 2009 [cited 2024 Jul 12]. p. 351–61. Available from: https://linkinghub.elsevier.com/retrieve/pii/B9780123747303000188.

[17] Hassan MA. Relational and NoSQL Databases: The Appropriate Database Model Choice. In: *2021 22nd International Arab Conference on Information Technology* [ACIT] [Internet]. Muscat, Oman: IEEE; 2021 [cited 2024 Jul 12]. p. 1–6. Available from: https://ieeexplore.ieee.org/document/9677042/.

[18] Inmon, W. H. What is a data warehouse?. *Prism Tech Topic*. 1995;[1[1]]:1–5.

[19] Gallo, C., De Bonis, M., Perilli, M. Data warehouse design and management: theory and practice. *IEEE Members Dipartimento di Scienze Economiche, Matematiche e Statistiche Università Di Foggia Largo Papa Giovanni Paolo II*. 2010;1–71121.

[20] Chen YA, Tripathi LP, Mizuguchi K. An integrative data analysis platform for gene set analysis and knowledge discovery in a data warehouse framework. *Database* [Internet]. 2016 [cited 2024 Jul 12];2016:baw009. Available from: https://academic.oup.com/database/article-lookup/doi/10.1093/database/baw009.

[21] Chen YA, Tripathi LP, Mizuguchi K. TargetMine, an Integrated Data Warehouse for Candidate Gene Prioritisation and Target Discovery. Uversky V, editor. *PLoS ONE* [Internet]. 2011 Mar 8 [cited 2024 Jul 12];6[3]:e17844. Available from: https://dx.plos.org/10.1371/journal.pone.0017844.

[22] Hu H, Correll M, Kvecher L, Osmond M, Clark J, Bekhash A, et al. DW4TR: A Data Warehouse for Translational Research. *Journal of Biomedical Informatics* [Internet]. 2011 Dec [cited 2024 Jul 12];44[6]:1004–19. Available from: https://linkinghub.elsevier.com/retrieve/pii/S1532046411001304.

[23] Lee GY, Alzamil L, Doskenov B, Termehchy A. A Survey on Data Cleaning Methods for Improved Machine Learning Model Performance [Internet]. arXiv; 2021 [cited 2024 Jul 12]. Available from: https://arxiv.org/abs/2109.07127.

[24] Neutatz, F, Chen, B, Abedjan, Z., Wu, E. From Cleaning before ML to Cleaning for ML. *IEEE Data Eng Bull* [Internet]. 2021;44[1], 24-41. Available from: https://

web.archive.org/web/20220407121053id_/http://sites.computer.org/debull/A21mar/p24.pdf.

[25] Cheng Z, Zou C, Dong J. Outlier detection using isolation forest and local outlier factor. In: *Proceedings of the Conference on Research in Adaptive and Convergent Systems [Internet]. Chongqing China: ACM;* 2019 [cited 2024 Jul 12]. p. 161–8. Available from: https://dl.acm.org/doi/10.1145/3338840.3355641.

[26] Hariri S, Kind MC, Brunner RJ. Extended Isolation Forest. *IEEE Trans Knowl Data Eng.* 2021 Apr 1;33[4]:1479–89.

[27] An W, Liang M, Liu H. An improved one-class support vector machine classifier for outlier detection. Proceedings of the Institution of Mechanical Engineers, *Part C: Journal of Mechanical Engineering Science* [Internet]. 2015 Feb [cited 2024 Jul 12];229[3]:580–8. Available from: http://journals.sagepub.com/doi/10.1177/0954406214537475.

[28] Amer M, Goldstein M, Abdennadher S. Enhancing one-class support vector machines for unsupervised anomaly detection. In: *Proceedings of the ACM SIGKDD Workshop on Outlier Detection and Description* [Internet]. Chicago Illinois: ACM; 2013 [cited 2024 Jul 12]. p. 8–15. Available from: https://dl.acm.org/doi/10.1145/2500853.2500857.

[29] Akbari Z, Unland R. Automated Determination of the Input Parameter of DBSCAN Based on Outlier Detection. In: *Iliadis L, Maglogiannis I, editors. Artificial Intelligence Applications and Innovations* [Internet]. Cham: Springer International Publishing; 2016 [cited 2024 Jul 12]. p. 280–91. [IFIP Advances in Information and Communication Technology; vol. 475]. Available from: http://link.springer.com/10.1007/978-3-319-44944-9_24.

[30] Alsini R, Almakrab A, Ibrahim A, Ma X. Improving the outlier detection method in concrete mix design by combining the isolation forest and local outlier factor. *Construction and Building Materials* [Internet]. 2021 Feb [cited 2024 Jul 12];270:121396. Available from: https://linkinghub.elsevier.com/retrieve/pii/S0950061820334000.

[31] De Goeij MCM, Van Diepen M, Jager KJ, Tripepi G, Zoccali C, Dekker FW. Multiple imputation: dealing with missing data. *Nephrology Dialysis Transplantation* [Internet]. 2013 Oct 1 [cited 2024 Jul 12];28[10]:2415–20. Available from: https://academic.oup.com/ndt/article-lookup/doi/10.1093/ndt/gft221.

[32] Sallaby AF, Azlan A. Analysis of Missing Value Imputation Application with K-Nearest Neighbor [K-NN] Algorithm in Dataset. *ijics* [Internet]. 2021 Aug 1 [cited 2024 Jul 12];5[2]:141. Available from: https://ejurnal.stmik-budidarma.ac.id/index.php/ijics/article/view/3185.

[33] Shah AD, Bartlett JW, Carpenter J, Nicholas O, Hemingway H. Comparison of Random Forest and Parametric Imputation Models for Imputing Missing Data Using MICE: A CALIBER Study. *American Journal of Epidemiology* [Internet]. 2014 Mar 15 [cited 2024 Jul 12];179[6]:764–74. Available from: https://academic.oup.com/aje/article-lookup/doi/10.1093/aje/kwt312.

[34] Hegde H, Shimpi N, Panny A, Glurich I, Christie P, Acharya A. MICE vs PPCA: Missing data imputation in healthcare. *Informatics in Medicine Unlocked.* 2019;17:100275.

[35] Feng HL, Fang YM, Xiang XQ, Li J, Li GH. A Data-Driven Noise Reduction Method and Its Application for the Enhancement of Stress Wave Signals. *The Scientific World Journal* [Internet]. 2012 [cited 2024 Jul 12];2012:1–7. Available from: http://www.hindawi.com/journals/tswj/2012/353081/.

[36] Xiong H, Gaurav Pandey, Steinbach M, Vipin Kumar. Enhancing data analysis with noise removal. *IEEE Trans Knowl Data Eng* [Internet]. 2006 Mar [cited 2024 Jul 12];18[3]:304–19. Available from: http://ieeexplore.ieee.org/document/1583581/.

[37] Ahuja A, Al-Zogbi L, Krieger A. Application of noise-reduction techniques to machine learning algorithms for breast cancer tumor identification. *Computers in Biology and Medicine* [Internet]. 2021 Aug [cited 2024 Jul 12];135:104576. Available from: https://linkinghub.elsevier.com/retrieve/pii/S001048252100370X.

[38] How to handle noise in machine learning [Internet]. Available from: I. https://www.geeksforgeeks.org/how-to-handle-noise-in-machine-learning/.

[39] Duboue P. The Art of Feature Engineering: Essentials for Machine Learning [Internet]. 1st ed. *Cambridge University Press*; 2020 [cited 2024 Jul 12]. Available from: https://www.cambridge.org/core/product/identifier/9781108671682/type/book.

[40] Feature engineering [Internet]. Available from: https://www.heavy.ai/technical-glossary/feature-engineering.

[41] Article on feature engineering [Internet]. Available from: https://builtin.com/articles/feature-engineering.

[42] Feature engineering [Internet]. Available from: https://en.wikipedia.org/wiki/Feature_engineering.

[43] Verdonck T, Baesens B, Óskarsdóttir M, Vanden Broucke S. Special issue on feature engineering editorial. *Mach Learn* [Internet]. 2024 Jul [cited 2024 Jul 12];113[7]:3917–28. Available from: https://link.springer.com/10.1007/s10994-021-06042-2.

[44] K. R. J, M. R, M. D, R. S, Gad A, K. J, et al. Feature optimization in high dimensional chemical space: statistical and data mining solutions. *BMC Res Notes* [Internet]. 2018 Dec [cited 2024 Jul 12];11[1]:463. Available from: https://bmcresnotes.biomedcentral.com/articles/10.1186/s13104-018-3535-y.

[45] Lemoine É, Dallaire F, Yadav R, Agarwal R, Kadoury S, Trudel D, et al. Feature engineering applied to intraoperative *in vivo* Raman spectroscopy sheds light on molecular processes in brain cancer: a retrospective study of 65 patients. *Analyst* [Internet]. 2019 [cited 2024 Jul 12];144[22]:6517–32. Available from: https://xlink.rsc.org/?DOI=C9AN01144G.

[46] Feinberg EN, Joshi E, Pande VS, Cheng AC. Improvement in ADMET Prediction with Multitask Deep Featurization. *J Med Chem* [Internet]. 2020 Aug 27 [cited 2024 Jul 12];63[16]:8835–48. Available from: https://pubs.acs.org/doi/10.1021/acs.jmedchem.9b02187.

[47] Lin C, Xiaoxiao Z. Optimizing Drug Screening with Machine Learning. In: *2022 19th International Computer Conference on Wavelet Active Media Technology and*

Information Processing [ICCWAMTIP] [Internet]. Chengdu, China: IEEE; 2022 [cited 2024 Jul 12]. p. 1–4. Available from: https://ieeexplore.ieee.org/document/10016572/.

[48] Fan W, Zhong E, Peng J, Verscheure O, Zhang K, Ren J, et al. Generalized and Heuristic-Free Feature Construction for Improved Accuracy. In: *Proceedings of the 2010 SIAM International Conference on Data Mining* [Internet]. Society for Industrial and Applied Mathematics; 2010 [cited 2024 Jul 12]. p. 629–40. Available from: https://epubs.siam.org/doi/10.1137/1.9781611972801.55.

[49] Zhao H, Sinha AP, Ge W. Effects of feature construction on classification performance: An empirical study in bank failure prediction. *Expert Systems with Applications* [Internet]. 2009 Mar [cited 2024 Jul 12];36[2]:2633–44. Available from: https://linkinghub.elsevier.com/retrieve/pii/S0957417408000444.

[50] Zhuang H, Wang X, Bendersky M, Najork M. Feature Transformation for Neural Ranking Models. In: *Proceedings of the 43rd International ACM SIGIR Conference on Research and Development in Information Retrieval* [Internet]. Virtual Event China: ACM; 2020 [cited 2024 Jul 12]. p. 1649–52. Available from: https://dl.acm.org/doi/10.1145/3397271.3401333.

[51] Galli S. Feature-engine: A Python package for feature engineering for machine learning. JOSS [Internet]. 2021 Sep 22 [cited 2024 Jul 12];6[65]:3642. Available from: https://joss.theoj.org/papers/10.21105/joss.03642.

[52] Patro SGK, Sahu KK. Normalization: A Preprocessing Stage. *International Advanced Research Journal in Science, Engineering and Technology* [Internet]. 2015 Mar 20 [cited 2024 Jul 12];20–2. Available from: http://www.iarjset.com/upload/2015/march-15/IARJSET%205.pdf.

[53] Starovoitov VV, Golub YuI. Data normalization in machine learning. *Informatika* [Minsk] [Internet]. 2021 Sep 30 [cited 2024 Jul 12];18[3]:83–96. Available from: https://inf.grid.by/jour/article/view/1156.

[54] What is data normalization [Internet]. Available from: https://www.geeksforgeeks.org/what-is-data-normalization/.

[55] Chen Z, Liu X, Zhao P, Li C, Wang Y, Li F, et al. *iFeatureOmega:* an integrative platform for engineering, visualization and analysis of features from molecular sequences, structural and ligand data sets. *Nucleic Acids Research* [Internet]. 2022 Jul 5 [cited 2024 Jul 12];50[W1]:W434–47. Available from: https://academic.oup.com/nar/article/50/W1/W434/6582173.

[56] Sabando MV, Ponzoni I, Milios EE, Soto AJ. Using molecular embeddings in QSAR modeling: does it make a difference? *Briefings in Bioinformatics* [Internet]. 2022 Jan 17 [cited 2024 Jul 12];23[1]:bbab365. Available from: https://academic.oup.com/bib/article/doi/10.1093/bib/bbab365/6366344.

[57] Krawczyk B. Learning from imbalanced data: open challenges and future directions. *Prog Artif Intell* [Internet]. 2016 Nov [cited 2024 Jul 12];5[4]:221–32. Available from: http://link.springer.com/10.1007/s13748-016-0094-0.

[58] Ebenezer, A. B, Boyinbode, O, Idowu, O. M. A Comprehensive Analysis of Handling Imbalanced Dataset. *IJATCSE*. 2021 Apr 5;10[2]:454–63.

[59] Chawla, Nitesh V, Bowyer, Kevin W, Hall, Lawrance O, Kegelmeyer, W. Philip. Chawla, Nitesh V.; Bowyer, Kevin W.; Hall, Lawrance O.; Kegelmeyer, W. Philip.

2002. SMOTE: Synthetic Minority. Journal of Artifical Intelligence Research. 2002;16:321–57.

[60] Fernandez A, Garcia S, Herrera F, Chawla NV. SMOTE for Learning from Imbalanced Data: Progress and Challenges, Marking the 15-year Anniversary. *jair* [Internet]. 2018 Apr 20 [cited 2024 Jul 12];61:863–905. Available from: http://jair.org/index.php/jair/article/view/11192.

[61] Blagus R, Lusa L. SMOTE for high-dimensional class-imbalanced data. BMC Bioinformatics [Internet]. 2013 Dec [cited 2024 Jul 12];14[1]:106. Available from: https://bmcbioinformatics.biomedcentral.com/articles/10.1186/1471-2105-14-106.

[62] Skryjomski, P, Krawczyk, B. Influence of minority class instance types on SMOTE imbalanced data oversampling. In *first international workshop on learning with imbalanced domains: theory and applications.* 2017 Oct;7–21.

[63] Haibo He, Yang Bai, Garcia EA, Shutao Li. ADASYN: Adaptive synthetic sampling approach for imbalanced learning. In: *2008 IEEE International Joint Conference on Neural Networks [IEEE World Congress on Computational Intelligence]* [Internet]. Hong Kong, China: IEEE; 2008 [cited 2024 Jul 12]. p. 1322–8. Available from: http://ieeexplore.ieee.org/document/4633969/.

[64] Dey I, Pratap V. A Comparative Study of SMOTE, Borderline-SMOTE, and ADASYN Oversampling Techniques using Different Classifiers. In: *2023 3rd International Conference on Smart Data Intelligence [ICSMDI] [Internet]. Trichy, India:* IEEE; 2023 [cited 2024 Jul 12]. p. 294–302. Available from: https://ieeexplore.ieee.org/document/10127891/.

[65] Gosain A, Sardana S. Handling class imbalance problem using oversampling techniques: A review. In: *2017 International Conference on Advances in Computing, Communications and Informatics [ICACCI]* [Internet]. Udupi: IEEE; 2017 [cited 2024 Jul 12]. p. 79–85. Available from: http://ieeexplore.ieee.org/document/8125820/.

[66] Sun Y, Wong AKC, Kamel MS. CLASSIFICATION OF IMBALANCED DATA: A REVIEW. *Int J Patt Recogn Artif Intell* [Internet]. 2009 Jun [cited 2024 Jul 12];23[04]:687–719. Available from: https://www.worldscientific.com/doi/abs/10.1142/S0218001409007326.

[67] Zhou Y, Liu Y. CORRELATION ANALYSIS OF PERFORMANCE METRICS FOR CLASSIFIER. In: *Decision Making and Soft Computing* [Internet]. João Pessoa [Paraíba], Brazil: WORLD SCIENTIFIC; 2014 [cited 2024 Jul 12]. p. 487–92. Available from: http://www.worldscientific.com/doi/abs/10.1142/9789814619998_0081.

[68] Liu Y, Zhou Y, Wen S, Tang C. A Strategy on Selecting Performance Metrics for Classifier Evaluation: *International Journal of Mobile Computing and Multimedia Communications* [Internet]. 2014 Oct 1 [cited 2024 Jul 12];6[4]:20–35. Available from: https://services.igi-global.com/resolvedoi/resolve.aspx?doi=10.4018/IJMCMC.2014100102.

[69] Japkowicz N. Assessment Metrics for Imbalanced Learning. In: He H, Ma Y, editors. *Imbalanced Learning* [Internet]. 1st ed. Wiley; 2013 [cited 2024 Jul 12]. p. 187–206. Available from: https://onlinelibrary.wiley.com/doi/10.1002/9781118646106.ch8.

[70] Caruana R, Niculescu-Mizil A. Data mining in metric space: an empirical analysis of supervised learning performance criteria. In: *Proceedings of the tenth ACM SIGKDD international conference on Knowledge discovery and data mining* [Internet]. Seattle WA USA: ACM; 2004 [cited 2024 Jul 12]. p. 69–78. Available from: https://dl.acm.org/doi/10.1145/1014052.1014063.

[71] Ferrer L. Analysis and Comparison of Classification Metrics [Internet]. *arXiv;* 2022 [cited 2024 Jul 12]. Available from: https://arxiv.org/abs/2209.05355.

[72] Davis J, Goadrich M. The relationship between Precision-Recall and ROC curves. In: *Proceedings of the 23rd international conference on Machine learning - ICML '06* [Internet]. Pittsburgh, Pennsylvania: ACM Press; 2006 [cited 2024 Jul 12]. p. 233–40. Available from: http://portal.acm.org/citation.cfm?doid=1143844.1143874.

[73] Hussin SK, Abdelmageid SM, Alkhalil A, Omar YM, Marie MI, Ramadan RA. Handling Imbalance Classification Virtual Screening Big Data Using Machine Learning Algorithms. Hassanien AEIB, editor. *Complexity* [Internet]. 2021 Jan 28 [cited 2024 Jul 12];2021:1–15. Available from: https://www.hindawi.com/journals/complexity/2021/6675279/.

Chapter 10

Transformative Drug Discovery: Comprehending the Impact of Generative Models on *De Novo* Drug Design

Anagha S. Setlur[1,*]
K. Chandrashekar[1]
Sunidhi[2]
and Vidya Niranjan[1,†]
[1]Department of Biotechnology, RV College of Engineering, Bangalore, India
[2]Centre of Excellence Computational Genomics, RV College of Engineering, Bangalore, India

Abstract

The pharmaceutical industry is undergoing a paradigm shift with the emergence of de novo drug design, fuelled by advancements in artificial intelligence (AI) and computational techniques. This chapter focuses on deep learning, reinforcement learning, and graph neural networks (GNNs), the cutting-edge generative de novo drug design models. These advanced models hold immense potential for effectively navigating vast chemical spaces, predicting novel drug-like molecules with desired biological activities, and optimizing drug candidates for superior efficacy and safety profiles.

The chapter delves into an advanced generative technique, including variational autoencoders (VAEs), generative adversarial networks (GANs), and others, to shed light on how they facilitate the creation of

* Corresponding Author's Email: anaghass@rvce.edu.in
† Corresponding Author's Email: vidya.n@rvce.edu.in

In: Harnessing AI for Drug Discovery
Editor: Kolla Bhanu Prakash
ISBN: 979-8-89530-540-9

entirely new drug candidates. The critical role of data preprocessing, molecular representation, and multi-omics data integration in enhancing the predictive power of these models is also comprehensively addressed. To illustrate the practical applications and transformative impact of these models in accelerating drug discovery, the chapter will incorporate real-world case studies and recent success stories. Furthermore, the discussion delves into this field's challenges and future directions. Key areas explored include improved interpretability of AI models, seamless integration with high-throughput screening methods, and the development of standardized benchmarks for robust model evaluation.

By providing a comprehensive overview of cutting-edge AI-driven generative models and their applications in de novo drug design, this chapter will equip researchers with a valuable resource to navigate this rapidly evolving field. Exploring challenges and future directions will guide further research efforts and pave the way for drug discovery's continued transformation through AI's power.

Keywords: De novo drug design, artificial intelligence, generative models, deep learning, reinforcement learning, graph neural networks, VAEs and GANs, transformative drug discovery

Introduction

Overview of De Novo Drug Design

Traditionally, drug discovery was mainly based on experimental techniques. However, a fundamental change began in the 1990s with the advent of computational approaches like de novo molecular design. De novo design is a computational method for generating novel chemical compounds without a preexisting template. This method attempts to automate the creation of new molecular structures with specific characteristics. It leverages knowledge derived from existing successful molecules to design novel ones with eccentric features.

In the context of drug discovery, de novo methods aim to generate molecules with unique properties that distinguish them from existing treatments and exhibit ideal drug-like qualities. This leads to challenges in many fields, as successful candidates must satisfy various pharmaceutical criteria beyond biological activity. These criteria include:

Target Selectivity: The designed molecule should notably interact with the specified target and avoid accidental interactions.

Favorable ADMET Profile: It is essential to optimize the compound's absorption, distribution, metabolism, excretion, and toxicity for effective drug delivery with minimal side effects [1].

Regardless of the innovative nature of de novo methods, several limitations in synthesizing desired molecules and the demand for computational techniques hindered their extensive adoption in drug discovery. Nevertheless, ongoing progression vows to conquer the challenges and further integrate de novo design into the drug discovery pipeline [2].

Evolution of Drug Discovery Methods

The hunt for medicinal remedies has a rich and ancient history, tracing back to the earliest days of human civilization. The history of drug discovery can be divided into three primary phases. In the 19th century, the evolution of medicine was primarily dependent on the use of plants and animals for medicinal purposes. They provided the raw materials for treatments to combat physical illnesses and as part of spiritual healing rituals. Afterward, the early 20th century played a critical role in discovering novel drug structures, recasting the antibiotic era. Finally, the late 20th century witnessed a significant boost in drug development through the advent of advanced technologies like high-throughput screening, combinatorial chemistry, and molecular modeling.

The development of recombinant DNA technology marked a turning point, making it possible to identify possible drug candidates [3].

The modern drug discovery and development period began in the early 20th century, with novel therapeutics often occurring within academic institutions during the early-stage exploration. The initial investigation is focused on detecting molecular mechanisms, disease-related macromolecules, and dysfunctional pathways. De novo drug design is laborious and complicated, with years of effort, considerable monetary investment, collaborative expertise across various disciplines, and significant dependence on data from innovative research strategies.

The initial investigation involves identifying a target for treatment, often a protein closely intertwined with the disease. Researchers examine the suitable targets that should expedite a response to medication and ultimately lead to improved health and reduced symptoms. Discovering the potential target is challenging, and drugs may also interact with unintended targets, leading to unpredictable effects on other cellular components.

After analyzing the target, the process is followed by discovering the potential therapeutic agents and the preclinical phase, wherein the experimental drugs are assessed in animal models to ensure safety and efficacy. Persistent progress aims to minimize or eliminate the dependence on animal testing.

If preclinical tests are positive, clinical trials are done on humans to assess the drug's safety and efficacy in patients and identify the candidates that exhibit an ideal benefit-to-risk ratio. The trial data is then proposed to regulatory authorities for in-depth evaluation and market authorization.

A successful regulatory evaluation leads to market authorization, enabling the drug to be sold and prescribed to patients. Nonetheless, the process continues. Post-market monitoring, often called pharmacovigilance studies or "phase 4" clinical trials, is crucial. This final stage involves continuous observation of patients to detect any side effects that might emerge with sustained use of the new treatment.

Automation significantly contributes to expediting the drug discovery and development process. Computer-assisted methods can automate many experiments and data analysis tasks, accelerating specific steps, fostering deeper knowledge acquisition, and minimizing errors [4, 5].

Role of AI in Modern Drug Discovery

Developing new drugs is a complex and expensive approach, with an average R&D cycle lasting 10-15 years and a high rate of failure. As a result, the pharmaceutical industry embraced artificial intelligence (AI) because of its potential to streamline the process and reduce costs.

AI offers several advantages, such as:

- Rapid identification of promising drug candidates
- Validation of drug targets
- Optimization of drug structure design

However, AI encounters challenges due to the vast and complex data involved, including its complexity, growth, diversity, and ambiguity. Despite these hurdles, AI tools like QSAR modeling and deep learning are effective in tasks such as:

- Predicting the desired chemical structure of a compound
- Screening large libraries of potential drug candidates
- Identifying sites for drug metabolism

Researchers are actively working on new AI methodologies like Deep Learning to improve the accuracy of predicting a compound's efficiency and side effects. Overall, AI holds a significant role in revolutionizing drug discovery by speeding up the process, reducing costs, and ultimately leading to the development of more effective treatments [6-8].

Fundamentals of Generative Models in Drug Design

Artificial Intelligence (AI) is gaining significant ground, with generative models experiencing considerable interest in recent years. The researcher's motivation to explore AI and generative models for designing de novo drugs has increased due to their diverse applications. The traditionally expensive drug design process can be significantly transformed by using Artificial Intelligence (AI). Entirely new biological compounds are created using generative models, which hold a particular promise for de novo drug design. However, the rapid pace of AI advancements and the built-in intricacy of drug design present a challenging environment for new researchers in this field. The investigation thoroughly examines the core principles adopted in generative models, including recurrent neural networks, autoencoders, generative adversarial networks, transformers, and hybrid models incorporating reinforcement learning [9].

The Variational Autoencoder (VAE) represents a robust class of generative models introduced by Diederik P. Kingma and Max Welling in 2013. VAEs are typically used to generate high-quality images without labeled data. They identify and condense the most important aspects of the input image. They use probability to represent the data, allowing for uncertainty and flexibility. The encoder converts the input image into a compressed representation (latent code) while the decoder reconstructs the original image from the latent code. The latent code is assumed to follow a normal distribution, thus maximizing the likelihood of accurate image reconstruction.

Variational Autoencoders (VAEs) convert molecular data into a continuous internal representation. This allows for gradient-based optimization of various molecular properties. This is an important advantage as it facilitates the generation of molecules tailored to possess specific target properties. It can preserve chemical similarity within the latent space. This allows for the generation of novel structures that are analogous to existing molecules. By systematically searching this latent space, researchers can identify novel compounds predicted to be active against specific targets. Researchers have used VAEs to generate drug-like molecules that meet five target properties within predefined ranges. They have also shown promise in adjusting individual properties without affecting others and even generating molecules that lie outside the training data distribution. Applying VAEs to extensive and intricate molecular datasets, including those containing natural products, offers the potential for a comprehensive exploration of the chemical space. This can lead to the generation of novel compounds with optimized functionalities. Natural products, while a rich source of bioactive compounds, pose a challenge due to their highly diverse structures, which VAEs can handle effectively [10].

Generative Adversarial Networks (GANs) were introduced in 2014 by Ian J. Goodfellow. They represent another dynamic generative model framework. There are two main components in GANs, unlike VAEs, namely the generator and the discriminator. These two components are involved in the adversarial training process, wherein the generator creates realistic synthetic data by studying the underlying patterns of input data. At the same time, the discriminator differentiates between the real and the generated data. The generator and the discriminator networks train simultaneously in a competitive dynamic to learn and replicate the patterns in the input data.

Variational Autoencoders (VAEs) use compressed latent representations to mimic input data and aim to generate outputs similar to the original data without using the labeled data. Conversely, Generative Adversarial Networks (GANs) employ supervised learning with a structured training process: the discriminator distinguishes accurate data from generated data, while the generator attempts to deceive the discriminator. This conflictual process produces more data and enhances complexity. VAEs have a smoother training process due to their exclusive objective, wherein GANs can make more complex and realistic data by capturing intricate details when well-trained. The competition between the discriminator and the generator enhances the diversity and realism of drug candidates. Conditional GANs (cGANs) can incorporate specific therapeutic properties during molecule creation. GANs excel at generating entirely new and previously unknown molecular structures. Their ability to discover fresh scaffolds—the fundamental chemical architectures of drugs—makes them essential.

These new scaffolds can serve as starting points for further optimization and development of entirely new classes of drugs [11].

Similar to VAEs, GANs can struggle with ensuring the generated molecules are novel and chemically feasible for synthesis in a lab setting. The quality of the training data plays a crucial role in GAN effectiveness. A diverse and well-curated dataset of known drugs is essential for the generator to learn the critical features of drug-like molecules. Fine-tuning the training process to balance the generator and discriminator properly is crucial for optimal results.

Overall, GANs offer a complementary approach to VAEs in de novo drug design. Their strength lies in generating exceptionally diverse and possibly unconventional drug-like molecules, pushing the boundaries of chemical space exploration [12].

Deep learning is a specialized domain of machine learning that uses artificial neural networks modeled after biological neurons. These networks are composed of interlinked units organized into an input layer and multiple hidden layers. Each unit processes information and transmits it to the subsequent units, thus learning to extract complex patterns by applying nonlinear transformations. Deep learning is effective in various machine learning applications such as computer vision, reinforcement learning, and natural language processing. Researchers also use it in supervised, unsupervised, and reinforcement learning frameworks.

Reinforcement learning is a technical area of machine learning dedicated to making optimal decisions to maximize reward in a given scenario. Researchers and practitioners use it in various applications and systems to determine the optimal course of action for specific circumstances. Reinforcement learning differs from supervised learning. It does not rely on predefined correct answers during training. Instead, the model learns through trial and error, guided by the reinforcement agent's decisions on achieving the task. Without a training dataset, it will inevitably learn from its experiences [13].

A branch of machine learning called reinforcement learning (RL) explores how an agent should make the best decisions possible in a changing environment. The agent engages with the environment and can change its status by doing so. Reward-based feedback is given to the agent, suggesting how desirable the final state is. Learning a policy or mapping from states to actions and maximizing the agent's predicted cumulative reward over time are the aims of reinforcement learning.

The critical elements of RL:

- *Agent:* The autonomous entity that learns to communicate with its surroundings to accomplish a specific objective is the essential component of reinforcement learning
- *Environment:* The external world with which an agent interacts. The agent receives feedback from this world in the form of rewards.
- *Action:* The set of options available to the agent.
- *Reward:* The information provided to the agent to assess the outcome of its actions.
- *Policy:* An agent's strategy for selecting actions based on the current situation.
- *Expected cumulative reward:* An agent's long-term goal is to maximize the total reward over time.

Reinforcement learning algorithms use an iterative approach to figure out optimal actions, thereby receiving feedback after each action, assessing the results, and adjusting the strategies. It's a useful method for automated systems that must make numerous minor judgments independently without human assistance.

In a self-governing, self-teaching system, reinforcement learning learns by making mistakes. It learns by doing to attain the best results, or, to put it another way, it takes actions to maximize rewards [14].

Graph Neural Networks (GNNs) constitute a specific class of artificial neural networks architected to handle data structured as graphs. GNNs operate on non-Euclidean data, expanding their applicability to diverse domains, unlike traditional neural networks, which process only Euclidean data (e.g., text, images). This document examines the core concepts of GNNs, their architectural characteristics, and practical applications in real-world situations. The fundamental principle associated with Graph Neural Networks (GNNs) is known as message passing to learn node representations. This iterative process enables GNNs to capture intricate relationships and patterns within the graph structure.

The message-passing process can be broken down into distinct stages:

- *Node Embeddings*: Each node in the graph is initially assigned a feature vector. This vector represents the inherent characteristics of the node.

- *Neighborhood Aggregation*: The GNN then aggregates information from the node's neighbors. This involves applying a specific function (e.g., mean, sum) to the feature vectors of neighboring nodes. This function combines the features of nearby nodes into a single representation.
- *Node Update*: The model incorporates the aggregated information, often the message, into the original feature vector of the central node. This process updates the node's representation, capturing the influence of its neighbors.
- *Graph Pooling:* In some applications, generating a single representation for the entire graph might be necessary. Another pooling function can aggregate the final node representations to achieve this [15].

Figure 1 highlights the generative drug discovery process.

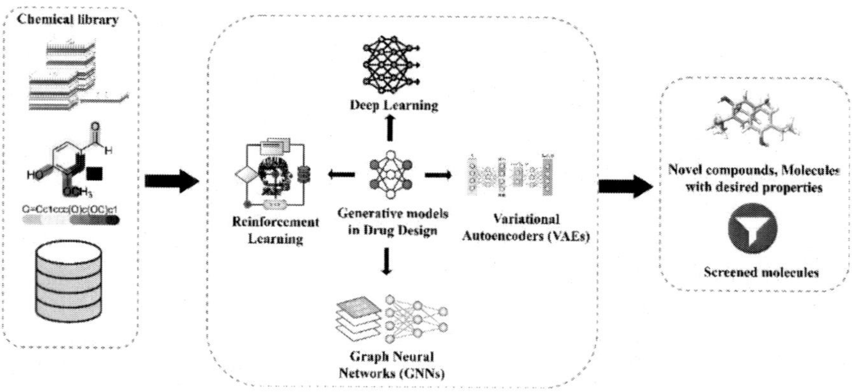

Figure 1. Illustration of fundamentals of generative models in drug design, showing the input variables such as the chemical library, process of transformative drug discovery using deep learning, reinforcement learning, VAEs, and GNNs, with the output being screened molecules, and identification of novel compounds.

Deep Learning for De Novo Drug Design

Drug development has greatly benefited from deep learning (DL), a potent artificial intelligence approach known for replicating human cognitive processes in areas notably image recognition and natural language processing. Using DL models to predict molecular features, such as affinity and toxicity, expedites drug development. New developments have expanded deep learning's capabilities to include de novo drug creation, creating innovative drug candidates by synergistically combining Reinforcement learning with deep learning.

Deep learning in drug discovery encompasses two primary components:

- A generative model, often a complex neural network, can generate novel molecular structures. This model acquires knowledge from existing drug data, such as SMILES strings or molecular graphs.
- A de novo drug design agent, functioning as a virtual chemist, manipulates generated molecules to enhance their properties. The generative model guides the agent's actions.

The process entails building a molecular structure step-by-step, considering the previously generated components and the desired attributes. The agent assesses the generated molecule and suggests modifications to optimize its fit for the target purpose (e.g., better binding or lower toxicity). This cycle continues until researchers identify a promising molecule candidate. Deep learning offers a revolutionary approach to drug discovery by facilitating the creation of entirely new drug molecules with desired properties [16, 17].

Applications in Drug Discovery

Traditionally, de novo drug design has involved constructing drug candidates by assembling molecular building blocks within a receptor pocket. Deep learning offers an alternative by leveraging its ability to learn known drugs' chemical and physical properties. A crucial step is extracting these relevant features, which is challenging due to the vast chemical search space and the nonlinear nature of drug optimization. Deep learning automates feature extraction, overcoming these challenges. Early attempts used string-based methods like SMILES generation models, but these have limitations in representing molecules accurately and conveying chemical properties. Molecular graph-based techniques utilizing graph neural networks have surpassed SMILES-based methods. Graph-based generation methods, however, can produce chemically incorrect intermediates. Researchers resolved this issue by investigating valid subgraphs and employing a junction tree approach. They explored several generative architectures, including RNNs, Autoencoders, Variational Autoencoders (VAEs), Adversarial Autoencoders (AAEs), and Generative Adversarial Networks (GANs). Researchers commonly use VAEs and GANs for molecular generation. VAEs encode molecules into a continuous latent space and decode them back into the molecular space. ChemVAE is a VAE-based SMILES generation model that can generate diverse molecules and estimate the distribution of encoded molecules. SD-VAE addresses syntactic and semantic issues in SMILES generation. GANs consist of a generator and a discriminator. The generator creates new data to deceive the discriminator, while the discriminator distinguishes real from fake data. Both components evolve during training. Objective-reinforced generative adversarial networks (ORGAN) combine GANs and reinforcement learning to train generative models for de novo drug design with specific properties. MolGAN uses GANs to generate diverse drug-like molecules, representing them using annotation matrices and dense adjacency tensors. Furthermore, L-Net can directly generate deep generative models of molecular graphs to generate high-quality three-dimensional (3D) drug-like compounds [18-20].

Case Study of Successful Deep Learning Models

Deep Learning Model for Kinase Inhibitors

- *Researchers:* RIKEN Centre for Computational Science (Japan) and Massachusetts Institute of Technology (MIT) teams
- *Specifics:* The team created a deep learning model trained on an enormous dataset of renowned kinase inhibitors and their distinctive features. In addition to predicting a molecule's binding affinity to a kinase target, the model, called "ACNN-DQN" (Attentional Convolutional Neural Network with Deep Q-Network), could suggest completely new drug-like structures with desirable features.
- *Results:* Researchers identified several interesting potential compounds for further studies. The study shows that deep learning can generate novel kinase inhibitors, although these haven't moved on to clinical trials yet [21].

Reinforcement Learning for Discovering MCR Ligands

- *Researchers:* A group from Mila, the Quebec Artificial Intelligence Institute, and the University of Toronto
- *Details:* This work used a deep reinforcement learning approach to discover novel ligands for the Mycobacterium tuberculosis mechanosensitive channel of small conductance (MscS). The concept concentrated on improving already-existing molecules rather than creating entirely new ones.
- *Results:* The model identified known MscS ligands and suggested several unique modifications to existing compounds that would improve their activity. These illustrations help us understand how deep learning optimizes leads in drug discovery [22].

These examples highlight the potential of deep learning models in de novo drug discovery, despite not yet representing perfect cases of medications reaching patients. They show how these models can be employed in designing novel drug-like structures, predicting the characteristics of possible drug candidates, and making the most of current leads to increased activity.

Reinforcement Learning

Generative models such as Variational Autoencoders (VAEs) or Generative Adversarial Networks (GANs) use reinforcement learning (RL) as their training method. The RL agent interacts with the generative model, offering rewards for creating molecules with desired properties, such as binding affinity to a target protein or drug-like characteristics. This reward-based feedback directs the generative model towards producing more promising drug candidates during training. Traditionally, drug design focuses on optimizing a single property at a time. However, RL allows for a more holistic approach. You can design the RL agent to consider multiple properties simultaneously, such as potency, selectivity, and pharmacokinetics (absorption, distribution, metabolism, and excretion). By maximizing a combined reward function that incorporates these factors, the RL agent guides the generation of molecules with a well-balanced profile suitable for drug development. This approach finds significant potential for discovering new drugs in a vast chemical universe. However, conventional techniques must frequently examine this area [14] adequately. RL can locate potential drug candidates proficiently by inspecting generative models and avoiding obstacles. Using the previously obtained data, RL agents prioritize areas with a greater chance of discovering promising therapeutic options. Combining RL with transfer learning helps the agent make valuable conclusions from sparse data, address a limited amount of data for novel targets, and speed up the design process. Iterative RL further optimizes candidates' drugs using the generative model, suggesting improvements in binding affinity and other vital characteristics by simulating drug-target interactions. These lead compounds are gradually optimized through successive refinement into viable drug candidates [23, 24].

Figure 2 emphasizes the various algorithms used for generative drug discovery using machine learning and deep learning.

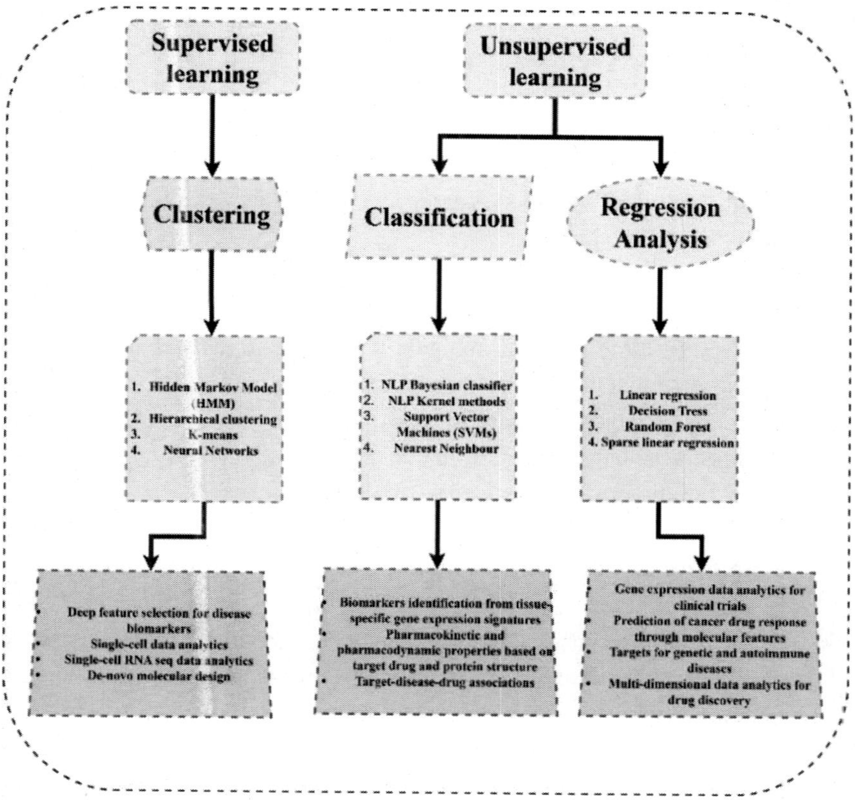

Figure 2. Algorithms for deep/machine learning in drug design, with applications.

Graph Neural Networks (GNNs) in Drug Design

Unlike conventional neural networks, which struggle to process graph-structured data, Graph Neural Networks (GNNs) are specifically designed to handle graph-based data. You can visualize molecules as networks, with atoms as nodes and chemical bonds as edges. GNNs excel at comprehending the spatial arrangements of atoms within a molecule and complex interactions, helping to understand the molecular properties and therapeutic applications. GNNs can adequately extract and encode atomic characteristics and inter-atomic relationships, enhancing tasks such as molecular generation, pattern detection related to drug-like behavior, and property prediction. They can model the interactions between a drug molecule and its target protein, predicting binding affinity and identifying possible alterations to improve the interactions by analyzing the representation of the chemical and the protein [15].

Representations of Molecular Structure

Graph neural networks (GNNs) represent molecular structures in drug development as graphs, where:

- *Nodes:* Represent the atoms within a molecule. Each node can store the type of atom (carbon, oxygen, nitrogen, etc.), its charge, and its hybridization state (sp^2, sp^3).
- *Edges*: Represent the bonds between the atoms in a molecule and can be unidirectional or bidirectional. They may also contain information about bond types such as single, double, or aromatic.
- *Atom Features:* Helps understand numerical representations of atomic properties such as atomic number, electronegativity, and hydrophobicity.
- *Bond Features:* They also help analyze numerical representations associated with edges, providing additional details like bond length or type.
- *Graph Construction:* Involves creating a network structure where atoms and bonds are nodes and edges (with or without features). The graph captures the spatial arrangement and connectivity of atoms within the molecule.

- *Message Passing:* In GNNs, neighboring nodes (atoms) exchange information along the graph's edges. This allows the network to learn how the properties of one atom are influenced by its surrounding atoms in the molecule [25, 26].

VAEs and Their Applications

Variational autoencoders (VAEs) play a critical role in de novo drug design by leveraging their ability to learn latent representations of molecules.

VAEs contribute to drug design in the following ways:

- *By generating novel molecules with desired properties:* To understand the chemical and physical features of present medications, VAEs could be trained on them. The model reduces this data into a latent space where every point denotes a possible molecule. Adjusting the latent space allows you to direct the decoders to produce unique compounds with specific favorable qualities, including binding affinity or decreased toxicity.
- *Efficient exploration of Chemical Space:* Traditional search methods are laborious due to the many potential medication candidates. With the help of VAEs, researchers may easily explore the latent space and look for the regions that match desired-property compounds. This investigation finds new therapeutic candidates that may go unnoticed through traditional means.
- *Data Augmentation for Drug Discovery*: VAEs can use preexisting datasets to produce novel and varied drug-like compounds. This enhanced data can train other machine learning models used in drug discovery activities. Predicting drug-target interactions or optimizing drug characteristics are two examples.
- *Understanding Drug-Target Interactions*: By investigating the implicit representations obtained by VAEs, scientists may develop information regarding the association between a molecule's composition and its physiological functions. It also helps understand how medications interact molecularly with their targets, which directs the development of more potent medications [27, 28].

Examples of VAE Applications

- *ChemVAE:* Analyses the distribution of encoded molecules in the latent space and generates SMILES strings, a typical representation of molecules. This enables researchers to create molecules with desired characteristics precisely.
- *SD-VAE*: The enhanced VAE, known as SD-VAE, overcomes the shortcomings of SMILES production by addressing syntactic and semantic problems and producing a more accurate representation of molecules [29].
- *GANs in drug design*: A discriminator and a generator are the two competing neural networks that contribute to forming a GAN. The purpose of the generator is to produce new compounds that mimic pharmaceuticals and can trick the discriminator into believing they are the real thing. In contrast, the discriminator functions as a critic, aiming to distinguish between the synthetic molecules produced by the generator and the actual molecules from a training dataset. The generator and discriminator both get enhanced functionality through an iterative training approach. The discriminator excels at locating fakes, while the generator learns to create increasingly realistic drug-like compounds. As a result of this continuous rivalry, more and more innovative and high-quality medication candidates are being developed [30, 31].

Advantages of GANs

- *Assessment of Diverse Chemical Space*: Unlike VAEs, GANs can navigate a dynamic latent space rather than a predetermined one. This allows them to explore a greater variety of chemical possibilities, developing more innovative and varied medication candidates.
- *Emphasise Realistic Molecules:* The adversarial character of GAN training encourages the generator to generate unique molecules that meet drug-like criteria. This is crucial for enhancing the chances that the molecules generated will be effective therapeutic candidates [32, 33].

Integration of Multi-Omics Data

Role of Multi-Omics Data in Drug Design

Traditionally, drug design has relied on a limited set of data points. Multi-omics data integration introduces a novel approach combining a comprehensive biological data array. This merges data from various sources, including proteomics, metabolomics, transcriptomics, and genomics, providing a broad spectrum of molecular diseases. By examining these interconnected datasets, researchers can discover new therapeutic targets, gain insights into disease progression mechanisms, and identify potential biomarkers for personalized medicine. In contrast, conventional methods often focus on specific targets, which can lead to issues in later stages of drug development. Multi-omics data enhances drug development by identifying disease-related networks and pathways, allowing researchers to target these complex structures more precisely. Furthermore, multi-omics predicts the drug's effectiveness and potential side effects by analyzing its impact on various biological processes, thus avoiding unforeseen safety issues and providing promising candidates for researchers in identifying medications with a high likelihood of success in clinical trials [34]. This helps identify specific genetic variations and molecular characteristics that affect drug response, thus supporting personalized medicines. Multi-omics data integration is a practical approach that has influenced the field of drug design. Multi-omics data has immense potential for accelerating the discovery and development of life-saving medications by enabling personalized therapy, improving target selection, and developing a deeper understanding of diseases [35, 36].

Techniques for Data Integration

Although multi-omics data is a wealth of information for drug discovery, its strength is the efficient integration of many disparate datasets.

Fundamental techniques used for multi-omics data integration in drug design:

a. *Similarity-Based Integration*: This approach identifies relationships between data points across different omics layers based on similarity.

The methods include:

- *Correlation analysis*: Analyses correlations between characteristics (such as gene expression and protein abundance) across datasets by measuring their linear relationships.
- *Network analysis*: Network analysis: To locate the underlying biological networks, it detects co-expressed genes, interacting proteins, or associated metabolites.

b. *Matrix-Based Integration*: This method uses matrices to represent data from different omics platforms.

The methods include:

- *Concatenation:* Concatenation is merely joining data matrices from several omics into one more extensive matrix. However, this approach could be less effective because data scales change naturally.
- *Matrix factorization*: An approach of reducing multiple omics matrices into a set of latent factors that outline underlying biological processes is known as matrix factorization. This makes it feasible to minimize dimensionality and find patterns identical to several datasets.

c. *Model-Based Integration*: The technique uses statistical or machine learning algorithms to integrate multi-omics data.

The methods include:

- *Canonical Correlation Analysis (CCA):* By identifying pairs of latent variables (one from each omics dataset) that have a significant correlation, CCA can determine the links between different kinds of data.
- *Multi-task learning*: It enables simultaneous training of a single model using features from different omics datasets to predict an expected outcome (e.g., drug response).

d. *Network-Based Integration*: Using integrated networks, this approach captures connections between several omics' layers' elements (genes, proteins, and metabolites). Researchers can use network analysis to find biomarkers, vital regulatory mechanisms, and potential therapeutic targets. Depending on the data type and biological interactions of interest, they can apply various network inference algorithms to construct these integrated networks [37, 38].

Enhancing Predictive Power with Multi-Omics Data

De novo drug design aims to create entirely new drug candidates, and multi-omics data integration significantly boosts its predictive power.

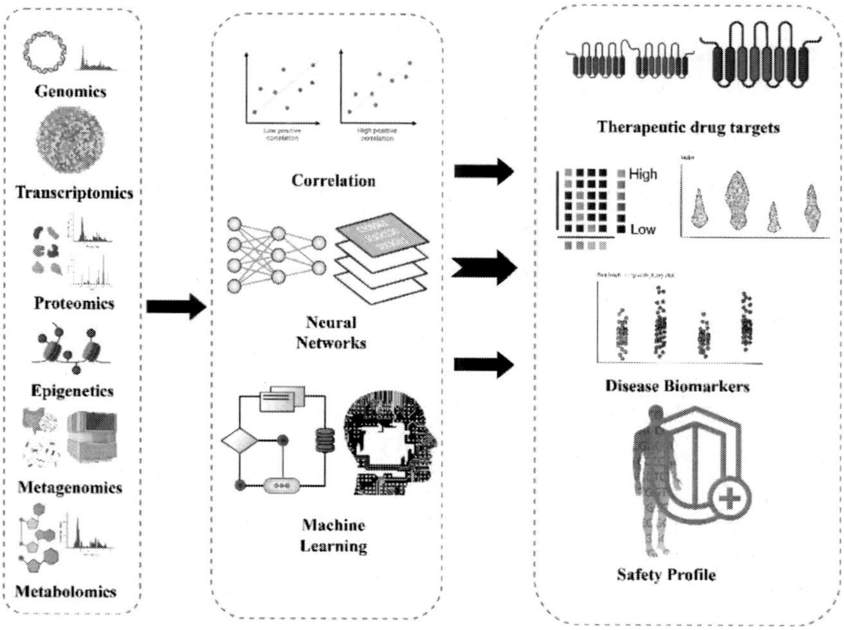

Figure 3. Integration of multi-omics data for enhanced drug discovery. A multi-omics layer is displayed, followed by data integration techniques, the predictive models, and the outcomes regarding therapeutic drug target identification, disease biomarker detection, or the development of a safety profile.

Traditional drug discovery approaches often focus on a limited number of targets due to reliance on small datasets. Multi-omics data provides a holistic view of disease by integrating proteins, metabolites, and genes, thus enabling researchers to identify critical regulatory points and novel drug targets. DL models such as GANs and VAEs adopt this data, which includes genetic variations, gene expression patterns, metabolic pathways, and interactions within the proteins, to predict the adverse effects and evaluate drug targets. This data prevents the need for extensive and costly clinical trials by predicting the drug's binding affinity and off-target effects by using in silico predictions. By examining how a medication candidate may interfere with metabolic processes or biological activities, multi-omics data might assist in identifying possible adverse effects. Complex datasets generated by multi-omics data lend themselves to machine learning approaches. Models for machine learning can be trained with the help of integrated multi-omics data by figuring out the connections between characteristics at various molecular levels and forecasting a novel drug candidate's safety and efficacy. Historical drug response data also helps by adding current knowledge regarding drug action; thus, the model's predictive ability can be further improved [35, 39, 40].

Figure 3 illustrates the role and integration of multi-omics data for enhanced drug discovery with various techniques and outcomes.

Challenges and Future Directions

Challenges in AI-Driven Drug Discovery

AI-driven drug discovery holds immense promise, but it's challenging. Here are some key hurdles researchers are working to overcome:

- *Data Availability and Quality*: De novo drug discovery requires high-quality data on current medications, the biology of disease, and molecular interactions. The amount of currently available data might need to be increased to train complicated AI models effectively. Data that is presently available might be unpredictable, noisy, or lacking from several sources. For accurate AI predictions, this data must be cleaned and standardized.
- *Interpretability and Explainability of AI Models:* Despite their strength, deep learning models' logic can be complex. Researchers must understand how an AI model determines a specific medication candidate to validate the findings and improve the design process.
- *Chemical Space Exploration*: Plenty of discontinuities in the potentially therapeutic space make it extremely broad. AI models

must be sufficiently sophisticated to navigate this complicated environment effectively and spot potential therapeutic leads that conventional approaches could overlook.

- *Ethical Considerations and Regulatory Frameworks*: The quality of AI models depends on the quality of the training data. Biased predictions might overlook therapeutic candidates that would benefit a specific patient population due to biases in the training data. Before clinical trials, AI-designed medications must pass stringent testing to guarantee both. Regulations may need to change to meet the difficulties AI-driven medication discovery poses.
- *Infrastructure and Computational Resources*: Sophisticated hardware and considerable quantities of processing power are necessary to train complex AI models for drug development. This might be a challenge for smaller research groups or universities [41, 42]

Need for Improved Model Interpretability

An essential requirement in AI-driven drug development is improved model interpretability. Although deep learning models are very effective, they must also be clarified. At times, we still determine how they arrive at their predictions about potential drugs. Improved interpretability enhances confidence in the outcomes and aids in well-informed decision-making by helping researchers comprehend the logic behind the model's recommendations. Without interpretability, scientists could be reluctant to follow the model's suggestions, which would slow the discovery process of new drugs. Researchers can learn more about which chemical properties are the most significant for the expected activity by understanding the logic behind the model. Using this information, researchers can cautiously design prospective medication candidates, emphasizing optimizing desired qualities and reducing undesirable consequences. Understanding how AI models arrive at their conclusions is crucial for developing more precise and effective drugs. Errors in the training data can affect the accuracy of AI models. These errors can only be identified and corrected if the model is transparent. This can lead to missing out on potentially beneficial drugs for specific patient groups. Before approving a drug for clinical trials, regulatory agencies need a clear understanding of how the drug works. Models that are difficult to interpret hinder the explanation of how an AI-designed drug functions. Enhanced interpretability facilitates communication with regulatory bodies and speeds up the approval process by offering a concise justification of the model's logic [21, 43].

Integration with High-Throughput Screening Methods

The combination of high-throughput screening (HTS) techniques and AI-driven drug discovery is a significant advancement in the field. HTS testing typically entails comparing hundreds or millions of molecules to a target molecule. This can be a costly and laborious operation. After being trained on large datasets, AI models with desired characteristics can be chosen as prospective therapeutic candidates. As a result, HTS tests a much smaller number of compounds, which accelerates and improves the efficacy of the screening process. It can be tricky to select the most promising candidates for further study because HTS can produce a lot of possible hits. Using AI models, data from HTS trials may be reviewed to determine which compounds have the best chance of success based on criteria like drug-likeness, projected activity, and other pertinent characteristics [44].

Development of Standardized Benchmarks

The potential for AI-driven drug discovery is vast, but establishing standardized benchmarks is crucial for advancement. It is only possible to compare the success of various AI models used for drug discovery activities with consistent benchmarks. Benchmarks provide a standard for evaluating a model's ability to find viable therapeutic options, predict drug-target interactions, or optimize pharmacological features. They also enable the sharing of data and models to ensure the reproducibility of results. This allows researchers to determine which models work best for particular jobs and monitor field progression. By enabling researchers to build on one another's work, standardized benchmarks facilitate transparency, trust, and advancement in this field [45].

Future Research Directions

Developing sophisticated deep-learning architectures for drug discovery is a significant challenge. To overcome this, models should be capable of identifying complex relationships between biological functions and molecular structures. Explainable AI (XAI) techniques enhance the understanding of model predictions and guide drug discovery efforts effectively, thereby enhancing model interpretability and reliability. Integrating diverse biological data into AI models, such as interactions within the proteins and metabolic pathways, deepens understanding and improves design strategies. Combined with AI models, a systems biology approach identifies novel drug targets and predicts potential adverse effects. AI and robotics can rationalize drug discovery processes, accelerating the synthesis and testing of AI-proposed drug candidates, improving efficiency, and reducing side effects. Ongoing ethical collaboration amongst researchers, ethicists, and regulatory bodies is essential to ensure ethical and responsible development and deployment of AI in drug discovery. Integrating AI, deep learning, and multi-omics data within ethical frameworks should address challenges such as AI bias and establish regulatory frameworks that protect patient safety while fostering innovation.

The successful combination of deep learning, artificial intelligence, and multi-omics data backed by solid moral and legal foundations could completely transform the drug discovery process [46].

Conclusion

The pharmaceutical industry is undergoing a seismic shift with the emergence of AI-driven drug discovery, offering an advanced, more rapid, and efficient data-centric approach to developing new medications. Traditional drug discovery methods rely on minimal data, but AI delves into multi-omics data such as proteomics, metabolomics, transcriptomics, and genomics. This is done to explore detailed disease-related structures and pathways, leading to novel therapeutic targets with enhanced efficiency. AI brings cutting-edge approaches using models like Variational Autoencoders (VAEs) and Generative Adversarial Networks (GANs), which may reduce the side effects by lessening the adverse effects. GANs, in particular, use a competitive approach to generate more refined drug designs wherein one network suggests drug structures while another assesses their effectiveness. In high-throughput screening (HTS), thousands of chemicals are compared to a target.

Declarations

Conflict of Interest

The authors declare no conflict of interest.

Author Statement

All authors agree to the publication of this article.

Acknowledgments

The authors thank RV College of Engineering's staff and non-staff members for aiding with resources, and the Centre of Excellence Computational Genomics interns, RVCE, for helping to collect data.

References

[1] Alam R, Rahman, G. S., Hasan, N., & Chowdhury, A. S. (2020). A De-Novo drug design and ADMET study to design small molecule stabilisers targeting mutant (V210I) human prion protein against familial Creutzfeldt-Jakob disease (fCJD). *International Journal of Computational Biology and Drug Design*, 13(1), 21-35.

[2] Crucitti D, Pérez Míguez, C., Díaz Arias, J. Á., Fernandez Prada, D. B., & Mosquera Orgueira, A. (2024). De novo drug design through artificial intelligence: an introduction. *Frontiers in hematology*, 3, 1305741.

[3] Pina AS, Hussain, A., & Roque, A. C. A. (2010). An historical overview of drug discovery. Ligand-Macromolecular Interactions in Drug Discovery: *Methods and Protocols,* 3-12.

[4] Moridani M, & Harirforoosh, S. (2014). Drug development and discovery: challenges and opportunities. *Drug discovery today*, 19(11), 1679-1681.

[5] Tang X, Dai, H., Knight, E., Wu, F., Li, Y., Li, T., & Gerstein, M. (2024). A Survey of Generative AI for De Novo Drug Design: New Frontiers in Molecule and Protein Generation. *ArXiv*, abs/2402.08703.

[6] Zhong F, Xing, J., Li, X., Liu, X., Fu, Z., Xiong, Z., ... & Jiang, H. (2018). Artificial intelligence in drug design. *Science China Life Sciences*, 61, 1191-1204.

[7] Floresta G, Zagni, C., Gentile, D., Patamia, V., & Rescifina, A. (2022). Artificial intelligence technologies for COVID-19 de novo drug design. *International journal of molecular sciences,* 23(6), 3261.

[8] Blanco-González A CA, Seco-González A, Conde-Torres D, Antelo-Riveiro P, Piñeiro Á, Garcia-Fandino R. The Role of AI in Drug Discovery: Challenges, Opportunities, and Strategies. *Pharmaceuticals* (Basel). 2023 Jun 18;16(6):891. doi: 10.3390/ph16060891. PMID: 37375838; PMCID: PMC10302890.

[9] Meyers J, Fabian, B., & Brown, N. (2021). De novo molecular design and generative models. *Drug discovery today*, 26(11), 2707-2715.

[10] Lim J, Ryu, S., Kim, J. W., & Kim, W. Y. (2018). Molecular generative model based on conditional variational autoencoder for de novo molecular design. *Journal of cheminformatics,* 10, 1-9.

[11] Lin E, Lin, C. H., & Lane, H. Y. (2020). Relevant applications of generative adversarial networks in drug design and discovery: molecular de novo design, dimensionality reduction, and de novo peptide and protein design. *Molecules*, 25(14), 3250.

[12] Bian Y, & Xie, X. Q. (2021). Generative chemistry: drug discovery with deep learning generative models. *Journal of Molecular Modeling*, 27, 1-18.

[13] Bai Q, Liu, S., Tian, Y., Xu, T., Banegas-Luna, A. J., Pérez-Sánchez, H., ... & Yao, X. (2022). Application advances of deep learning methods for de novo drug design and molecular dynamics simulation. Wiley Interdisciplinary Reviews: *Computational Molecular Science,* 12(3), e1581.

[14] Krishnan SR, Bung, N., Bulusu, G., & Roy, A. (2021). Accelerating de novo drug design against novel proteins using deep learning. *Journal of Chemical Information and Modeling,* 61(2), 621-630.

[15] Satorras VG, Hoogeboom, E., & Welling, M. (2021, July). E (n) equivariant graph neural networks. In *International conference on machine learning* (pp. 9323-9332). PMLR.

[16] Xiong J, Xiong, Z., Chen, K., Jiang, H., & Zheng, M. (2021). Graph neural networks for automated de novo drug design. *Drug discovery today*, 26(6), 1382-1393.

[17] Ryu JY, Lee, M. Y., Lee, J. H., Lee, B. H., & Oh, K. S. (2020). DeepHIT: a deep learning framework for prediction of hERG-induced cardiotoxicity. *Bioinformatics,* 36(10), 3049-3055.

[18] Isert C, Atz, K., & Schneider, G. (2023). Structure-based drug design with geometric deep learning. *Current Opinion in Structural Biology*, 79, 102548.

[19] Chen H, Engkvist, O., Wang, Y., Olivecrona, M., & Blaschke, T. (2018). The rise of deep learning in drug discovery. *Drug discovery today*, 23(6), 1241-1250.

[20] Abe K, Ozako, M., Inukai, M., Matsuyuki, Y., Kitayama, S., Kanai, C., ... & Yoshimori, A. (2023). Deep learning driven de novo drug design based on gastric proton pump structures. *Communications Biology*, 6(1), 956.

[21] Mouchlis VD, Afantitis, A., Serra, A., Fratello, M., Papadiamantis, A. G., Aidinis, V., ... & Melagraki, G. (2021). Advances in de novo drug design: from conventional

to machine learning methods. *International journal of molecular sciences*, 22(4), 1676.

[22] Courtens C, Risseeuw, M., Caljon, G., Cos, P., & Van Calenbergh, S. (2018). Acyloxybenzyl and alkoxyalkyl prodrugs of a fosmidomycin surrogate as antimalarial and antitubercular agents. *ACS medicinal chemistry letters*, 9(10), 986-989.

[23] Fang Y, Pan, X., & Shen, H. B. (2023). De novo drug design by iterative multiobjective deep reinforcement learning with graph-based molecular quality assessment. *Bioinformatics,* 39(4), btad157.

[24] Gummesson Svensson H, Tyrchan, C., Engkvist, O., & Haghir Chehreghani, M. (2024). Utilizing reinforcement learning for de novo drug design. *Machine Learning,* 113(7), 4811-4843.

[25] Klicpera J, Becker, F., & Günnemann, S. (2021, December). Gemnet: Universal directional graph neural networks for molecules. In *Proceedings of the 35th International Conference on Neural Information Processing Systems* (pp. 6790-6802).

[26] Scarselli F, Gori, M., Tsoi, A. C., Hagenbuchner, M., & Monfardini, G. (2008). The graph neural network model. *IEEE transactions on neural networks*, 20(1), 61-80.

[27] Ilnicka A, & Schneider, G. (2023). Designing molecules with autoencoder networks. *Nature Computational Science,* 3(11), 922-933.

[28] Munson BP, Chen, M., Bogosian, A., Kreisberg, J. F., Licon, K., Abagyan, R., ... & Ideker, T. (2024). De novo generation of multi-target compounds using deep generative chemistry. *Nature Communications*, 15(1), 3636.

[29] Simonovsky M, & Komodakis, N. (2018). Graphvae: Towards generation of small graphs using variational autoencoders. In *Artificial Neural Networks and Machine Learning–ICANN 2018: 27th International Conference on Artificial Neural Networks*, Rhodes, Greece, October 4-7, 2018, Proceedings, Part I 27 (pp. 412-422). Springer International Publishing.

[30] Prykhodko O, Johansson, S. V., Kotsias, P. C., Arús-Pous, J., Bjerrum, E. J., Engkvist, O., & Chen, H. (2019). A de novo molecular generation method using latent vector based generative adversarial network. *Journal of cheminformatics*, 11, 1-13.

[31] Rossetto AM, & Zhou, W. (2019, September). Gandalf: A prototype of a gan-based peptide design method. In *Proceedings of the 10th ACM international conference on bioinformatics, computational biology and health informatics* (pp. 61-66).

[32] Anand N, & Huang, P. (2018). Generative modeling for protein structures. *Advances in neural information processing systems*, 31.

[33] Karimi M, Zhu, S., Cao, Y., & Shen, Y. (2020). De novo protein design for novel folds using guided conditional wasserstein generative adversarial networks. *Journal of chemical information and modeling*, 60(12), 5667-5681.

[34] Mohammadi-Shemirani P, Sood, T., & Paré, G. (2023). From 'omics to multi-omics technologies: the discovery of novel causal mediators. *Current atherosclerosis reports,* 25(2), 55-65.

[35] Paananen J, & Fortino, V. (2020). An omics perspective on drug target discovery platforms. *Briefings in bioinformatics*, 21(6), 1937-1953.

[36] Vahabi N, & Michailidis, G. (2022). Unsupervised multi-omics data integration methods: a comprehensive review. *Frontiers in genetics*, 13, 854752.

[37] Graw S, Chappell, K., Washam, C. L., Gies, A., Bird, J., Robeson, M. S., & Byrum, S. D. (2021). Multi-omics data integration considerations and study design for biological systems and disease. *Molecular omics*, 17(2), 170-185.

[38] Athieniti E, & Spyrou, G. M. (2023). A guide to multi-omics data collection and integration for translational medicine. *Computational and Structural Biotechnology Journal*, 21, 134-149.

[39] Subramanian I, Verma, S., Kumar, S., Jere, A., & Anamika, K. (2020). Multi-omics data integration, interpretation, and its application. *Bioinformatics and biology insights*, 14, 1177932219899051.

[40] Ivanisevic T, & Sewduth, R. N. (2023). Multi-omics integration for the design of novel therapies and the identification of novel biomarkers. *Proteomes,* 11(4), 34.

[41] Blanco-Gonzalez A, Cabezon, A., Seco-Gonzalez, A., Conde-Torres, D., Antelo-Riveiro, P., Pineiro, A., & Garcia-Fandino, R. (2023). The role of AI in drug discovery: challenges, opportunities, and strategies. *Pharmaceuticals,* 16(6), 891.

[42] Hughes JP, Rees, S., Kalindjian, S. B., & Philpott, K. L. (2011). Principles of early drug discovery. *British journal of pharmacology*, 162(6), 1239-1249.

[43] Mak KK, Wong, Y. H., & Pichika, M. R. (2024). Artificial intelligence in drug discovery and development. *Drug discovery and evaluation: safety and pharmacokinetic assays*, 1461-1498.

[44] Zhuang D, & Ibrahim, A. K. (2021). Deep learning for drug discovery: a study of identifying high efficacy drug compounds using a cascade transfer learning approach. *Applied Sciences*, 11(17), 7772.

[45] Bannigan P, Aldeghi, M., Bao, Z., Häse, F., Aspuru-Guzik, A., & Allen, C. (2021). Machine learning directed drug formulation development. *Advanced Drug Delivery Reviews,* 175, 113806.

[46] Pérez Santín E, Rodríguez Solana, R., González García, M., García Suárez, M. D. M., Blanco Díaz, G. D., Cima Cabal, M. D., ... & López Sánchez, J. I. (2021). Toxicity prediction based on artificial intelligence: A multidisciplinary overview. Wiley Interdisciplinary Reviews: *Computational Molecular Science,* 11(5), e1516.

Chapter 11

Leveraging Artificial Intelligence for Data Collection and Analysis in Drug Discovery

Preetha Mary George[1,*]
G. Soniya Priyatharsini[2]
and N. S. Shubhashree[1]
[1]Department of Physics, [2]Department of Computer Science and Engineering,
Dr. M. G. R. Educational and Research Institute of Technology, Maduravoyal,
Chennai, Tamil Nadu, India

Abstract

Artificial Intelligence (AI) has ushered in a progressing era in drug discovery, revolutionizing traditional methodologies and accelerating the pace of innovation. This paper explores the pivotal role of AI techniques across various stages of drug discovery, from data collection to predictive modeling and target interaction analysis. Key data resources such as CHEMBL and Drug Bank are invaluable repositories, providing essential chemical and biological information. Leveraging AI algorithms, these vast datasets are converted into computationally accessible formats, enabling sophisticated analyses. AI techniques are crucial in predictingdrug properties such as toxicity, physiochemical attributes, and bioactivity. By harnessing Machine Learning (ML) and Deep Learning (DL) algorithms, researchers may rapidly assess all the potential of candidate compounds, streamlining drug development in the early stage. The design of novel molecules in drug are streamlined by the new AI driven models. The target to drug interactions are explored and enabled. Researchers make decisions based on the AI suggestions and the

* Corresponding Author's Email: preethamarygeroge.phy@drmgrdu.ac.in

In: Harnessing AI for Drug Discovery
Editor: Kolla Bhanu Prakash
ISBN: 979-8-89530-540-9

mechanisms of binding are targeted for predicting structures. The decisions consider the poses and several challenges. This includes the considerations of ethics and the understandability of the model. This secures the therapies. This usage gives constructive AI solutions. There are many opportunities in the drug discovery field. There are new revolutions happening in drug discovery, which is getting pipelined. The consolidations of the data where the information is collected may vary with the sources. By easing the combination of the pharma companies and the AI expert systems the regulatory bodies leverage the unmet needs of the medical field. It improves the outcome of the healthcare globally.

Keywords: drug, machine learning, deep learning, artificial intelligence, bioinformatics, data resources

Introduction

By collaborating the objective models, exploratory and summary are the new possible remedial entries involved in drug discovery. Despite the development of the systems and technologies, the drug discovery procedure remains challenging [1, 2]. It is an interesting and efficient process to discover drugs in this modern world. It is because of the biological target. The shape of the water and charge are the important aspects where they interact and bind duringthe design of molecules. Computer modelling techniques and bioinformatics interact and bind. Nowadays special methods are used for designing purposes. They are represented by the progressive of the majority drugs. This will happen with the small particles.

Research on the modelling animals and cell base are diagnosed in the drug development and discovery. The step-by-step process is approved by the regulations after interviewing the humans by clinical trials. The newest forms of identifying the chemistry to their harmony with precision and solubility. All of these compounds is identified and proceed with the drug development in the clinical trials.

Resources and Methods for Drug Discovery

For the evaluation of the Drug discovery, there are three components were used. Those are Data resources, Data representation schemes and AI methods. They are explained in the following Table 1.

Data Resources

Table 1. Data resources and its uses

Sl. No	Data resources	Uses
1.	ChEMBL	A curated database of bioactive molecules with drug-like properties, facilitating the translation of genomic data into effective new drugs.
2.	ChemDB	A chemical database containing over 5 million commercially available small molecules with predicted or experimentally determined physicochemical properties.
3.	COCONUT	A comprehensive database of over 400,000 natural products with detailed molecular properties and descriptors.
4.	DGIdb	Provides information on drug-gene interactions and druggable genomes from multiple trusted sources.
5.	DrugBank	A database containing detailed information about drugs, their targets, 3D structures, and other properties.
6.	DTC	Offers bioactivity data and classification of drug targets
7.	INPUT	A platform providing information on traditional Chinese medicine, including compounds from various herbs
8.	PubChem	An open chemistry database offering extensive chemical information, including molecular structures, identifiers, physical properties, and biological activities.
9.	SIDER	A database that provides information on marketed drugs and their adverse effects.
10.	STITCH	A database that predicts interactions between chemicals and proteins, covering a wide range of organisms.

Molecular Descriptors and Representation

Molecular descriptors play a crucial role in AI-based drug discovery, enabling the analysis and representation of molecular structures [4-8]. These descriptors, which can be categorized into various dimensions, help retain the intrinsic physicochemical properties of substances in a computer-readable format. Open-source toolkits like Open Babel and ChemmineR are available for calculating molecular descriptors and structural representations. Figure 1 explains the above in detail.

Zero-Dimensional (0D) Descriptors: The simplest form, derived from the chemical formula, including molecular weight, atomic number, and atom-type count.

(1D) One dimensional encodes the atom fragments, functional groups, and the rings. ID descriptor, SMILES (simplified molecular input line entry) is usually used.

(2D) Two-dimensional descriptors can be built on by incorporating connectivity, topological features and adjacency. One of the 2d descriptors is molecular 2d fingerprints. They are represented by edges and atoms.

(3D) Three dimensional descriptors uses the molecules in configurations, surface area, special and geometric and volume configurations etc.

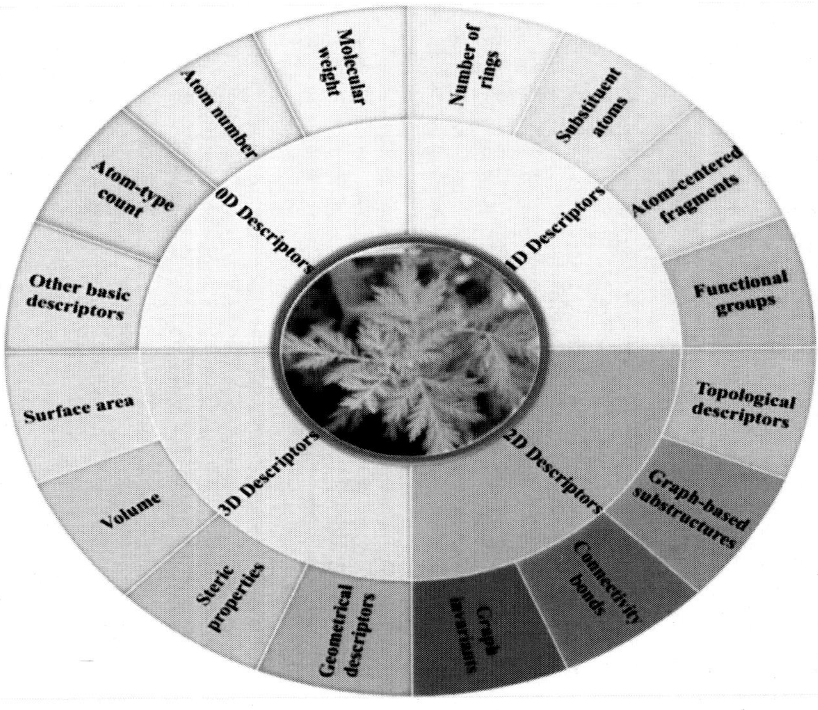

Figure 1. Summary of molecular and structural representation schemes.

AI Techniques Used

These innovative techniques improvised the discovery in drug. These traditional techniques are expensive and time consuming [9]. Such techniques are included in the unsupervised and supervised learning techniques. They are used in the drug discovery. The following figure explains the techniques and the methods in the Table 2.

- *Supervised Learning*: Utilizes labelled data to train models for classification and prediction tasks. Standard methods include Support

Vector Machines (SVM), Convolutional Neural Networks (CNN) [10]. Recurrent Neural Networks (RNN), and Generative Adversarial Networks (GAN).

- *Unsupervised Learning*: Works with unlabeled data to identify patterns and clusters without prior knowledge. Techniques include K-means clustering, hierarchical clustering, and dimensionality reduction methods like Principal Component Analysis (PCA) and t-distributed Stochastic Neighbour Embedding (t-SNE).

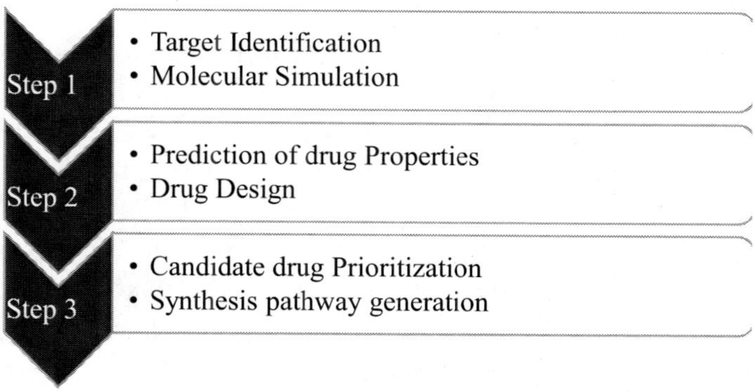

Figure 2. AI Techniques used in the drug discovery.

Table 2. Methods and Techniques involved in the AI Techniques

Category	Task	Method
Supervised learning	Regression analysis	SVM (Support Vector Machine) CNN (Convolution Neural Network) RNN (Recurrent Neural Network) GAN (Generative adversarial Network)
Unsupervised learning	Clustering	1. K-means clustering 2. Hierarchical clustering
	Dimension Reduction	PCA t -SNE

Regression Analysis in Drug Discovery

Regression analysis [11] is a fundamental statistical method used to understand and predict the relationships between variables. Regression models are employed in drug discovery to explore how different molecular descriptors or experimental conditions influence biological activities or drug efficacy.

- *Multiple Linear Re*gression (MLR): [12] This technique is commonly used to model the relationship between a dependent variable (e.g., drug activity) and multiple independent variables (e.g., molecular descriptors). By fitting a linear equation to the observed data, MLR can predict outcomes for new data points. The model determines the best-fit line by minimizing the sum of squared errors—the differences between observed values and those predicted by the model. MLR is particularly useful when the relationship between variables is assumed to be linear, but it may struggle with non-linear relationships common in complex biological systems.
- *Decision Tree (DT):* [13] Decision trees are a non-linear supervised learning technique that can be applied to classification and regression tasks. DTs are valuable for handling complex, non-linear relationships between variables in drug discovery. The DT algorithm starts at the root node and selects a branch based on the decision rule associated with that node. The process continues through internal nodes, making decisions at each step until it reaches a leaf node representing the predicted outcome. Decision trees are intuitive and easy to interpret, making them popular for modeling decision-making processes in drug design.
- *Logistic Regression (LR):* [14] Logistic regression is used to model the probability of a binary outcome (e.g., the presence or absence of a biological activity) based on one or more predictor variables. Unlike linear regression, which predicts a continuous outcome, LR estimates the probability that a given input belongs to a particular category. This is particularly useful in drug discovery for predicting whether a compound will be active or inactive against a specific target. LR can be extended to handle multiple classes (multinomial logistic regression) or ordered outcomes (ordinal logistic regression).
- *Support Vector Machine (SVM):* [15] SVM is a powerful supervised learning technique in classification and regression tasks. In drug discovery, SVMs are often applied to classify compounds based on their biological activity. SVM works by mapping the input data into a higher-dimensional feature space using kernel functions and finding the optimal hyperplane separating the different classes with maximum margin. This technique is particularly effective for high-dimensional datasets, such as those encountered in cheminformatics and bioinformatics.

- *Convolutional Neural Networks (CNN):* [16] CNNs are a type of deep learning model particularly suited for processing structured grid-like data, such as images or molecular graphs. In drug discovery, CNNs can automatically extract features from molecular structures and predict biological activity. The architecture of a CNN typically includes convolutional layers that perform feature extraction, pooling layers that reduce the dimensionality of the data, and fully connected layers that output the final prediction. CNNs are advantageous in drug discovery because they can learn complex, non-linear relationships without requiring manual feature engineering.
- *Recurrent Neural Networks (RNN):* [17] RNNs are designed to handle sequential data, making them suitable for tasks where the order of inputs is essential, such as predicting the properties of a molecule based on its SMILES string. RNNs contain loops that allow information to persist across steps in the sequence, enabling the model to capture temporal dependencies. In drug discovery, RNNs have been applied to generate novel chemical structures, predict drug-target interactions, and model time-series data from biological experiments.
- *Generative Adversarial Networks (GAN):* [18] GANs consist of two neural networks—the generator and the discriminator—trained simultaneously in a competitive framework. The generator creates new data samples, while the discriminator evaluates whether the samples are actual or generated. GANs can generate novel chemical compounds with desired properties in drug discovery by learning from existing drug datasets. The ability of GANs to produce high-quality, diverse molecules makes them a promising tool for de novo drug design.

Clustering Techniques in Drug Discovery

Clustering is an unsupervised learning technique that groups similar data points into clusters based on their features [19-25]. In drug discovery, clustering is used to identify compounds with similar chemical properties or biological activities, which can help prioritize candidates for further development.

- *K-means Clustering:* K-means is one of the most widely used clustering algorithms in drug discovery. It works by partitioning the dataset into K clusters and assigning each data point to the nearest centroid. The centroids are the mean of the data points in each cluster. The algorithm iteratively adjusts the centroids and reassigns data points until the cluster assignments stabilize. K-means is effective for discovering natural groupings in data, such as grouping compounds with similar pharmacophores or identifying subpopulations in high-throughput screening data.
- *Hierarchical Clustering:* Hierarchical clustering builds a tree-like structure (dendrogram) to represent the nested grouping of data points. It can be performed in agglomerative (bottom-up) or divisive (top-down). In agglomerative clustering, each data point starts as its cluster, and pairs of clusters are merged iteratively based on their similarity until a single cluster remains. Divisive clustering starts with all data points in one cluster and splits them recursively. Hierarchical clustering is particularly useful in drug discovery for analyzing the structural similarity of compounds and organizing them into a hierarchy that reflects their chemical relationships.
- *Principal Component Analysis (PCA):* PCA is a dimensionality reduction technique that transforms a large set of variables into a smaller one that still contains most information. PCA is often used before clustering to reduce the complexity of the data, making it easier to visualize and cluster. In drug discovery, PCA helps to identify the essential features that contribute to the variation in the dataset, which can then be used to cluster compounds or analyze their distribution in chemical space.
- *t-Distributed Stochastic Neighbour Embedding (t-SNE):* t-SNE is a non-linear dimensionality reduction technique for visualizing high-dimensional data in two or three dimensions. It works by converting the similarities between data points into probabilities and minimizing the divergence between them in high- and low-dimensional spaces. t-SNE is particularly useful in drug discovery for visualizing the clustering of compounds based on their chemical properties or biological activities, allowing researchers to identify patterns and relationships that may not be apparent in the original high-dimensional data.

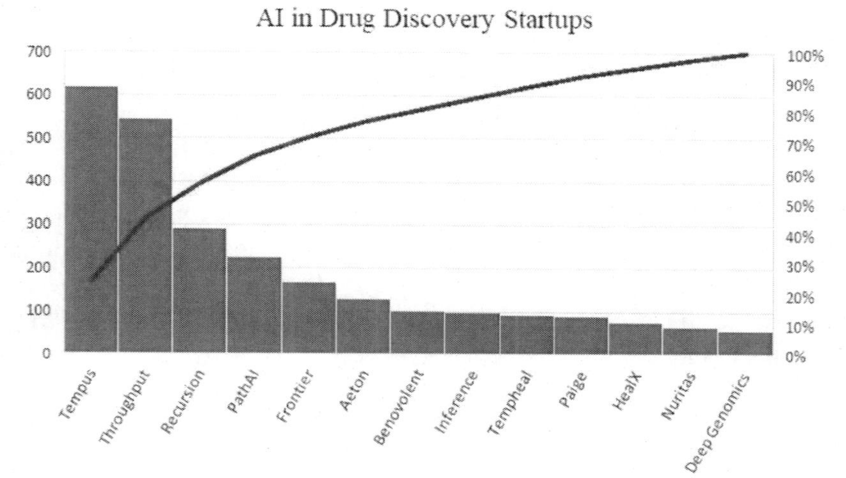

Figure 3. List of the Startups concentrated on the Drug Discovery using AI.

Classification of the drug discovery based on the startups is explained in Figure 3.

Advanced Applications of AI in Drug Design

AI has advanced beyond traditional applications and is now playing a critical role in designing and discovering new drugs [26-30]. These advanced applications include designing novel drug molecules, predicting protein structures, and exploring complex drug-target interactions.

- *Designing De novo Drugs*: De novo drug design involves generating new chemical entities from scratch without relying on existing templates. AI techniques and intense learning models have revolutionized this process by enabling the rapid generation of novel drug-like compounds. Models like ReLeaSE (Reinforcement Learning for Structural Evolution), ChemVAE (Variational Autoencoder), Graph INVENT, and MolRNN (Molecular Recurrent Neural Network) are used to design new molecules. These models use SMILES strings, molecular graphs, and 3D geometries to generate structures with optimized properties, such as binding affinity, solubility, and bioavailability.

- *Prediction of drug target interactions*: There are many essential interactions for the understanding in the drug biological targets. This AI models uses strong and severe approaches in learning for the biological targets. There are DTI predictions for the developed models. This uses the information to be like the prediction handling model. Information that are deep in belief networks, multi layer perception and CNN are the commonly used DTI predictions. The models which are used by the AI prediction models are used for the identification of the drug discovery. These effects predict the off target effects. It also repurposes the existing techniques in discovering the drug.
- *Prediction of the antagonism and drug synergism*: This process involves the combinations of the antagonistic effects and synergistic effects. This includes reducing each other's efficacy and enhancing this efficacy.

Predictions of these drug discoveries is effective in these combinations. The therapies are complex in treating severe diseases like cancer, HIV and so on. Bayesian algorithm, random forest algorithm and so on are used to predict the chemical genetic interaction data. Some of the deep learning models like deep synergy used to integrate information about diseases and their synergistic effects. These predictions help the researchers to predict the drug discovery which is used in the clinical trials.

- *Protein Structures Prediction*: This prediction of the structure of proteins with the crucial functions of the drug designs. Methods that include AI designs like alpha fold which are developed by the important strides area. Newly developed technological platforms like alpha fold predict the amino acid from the sequence. This gives the accuracy level of the comparable experimental methods which is like crypto electron microscopy and x-ray crystallography. This importance of the developments in the advancements and the negotiations handling for the discovery of the drug are targeted by the acceleration of the structural information. This is used in the structure-based design and virtual screen design.
- *Drug test binding affinity prediction*: During the prediction of the DTI, it usually focuses on the drug that will bind the target. This prediction explains the detailed information about the understanding

of affinity binding. The strength of the interaction is based on the detailed prediction. Drug efficacy can be measured by the dissociation and the constants inhibition. Eventhough these measurements are costly and time-consuming; researchers are suggesting new model predictions. Innovative models like deep data which is predicted with the binding affinity. this is done by decoding the smiles techniques. The compounds are prioritised about the future estimating potential efficacy.

- *Designing Nanomedicines using AI*: The upcoming new technology is opening new avenues. In the drug discovery the nanomedicines are designed in the target-specific cells. The tissues are specific cells that are optimised by the compositions. AI models were used in the prediction of biological systems. This creates new formulations. For example, the improvised new AI models are predicted in the cellular nanoparticles. This research explains the enhanced efficacy and the reduction of toxicity. Printing technologies like 3D are explored for creating the physical and chemical properties, which is customised with nanomedicines.

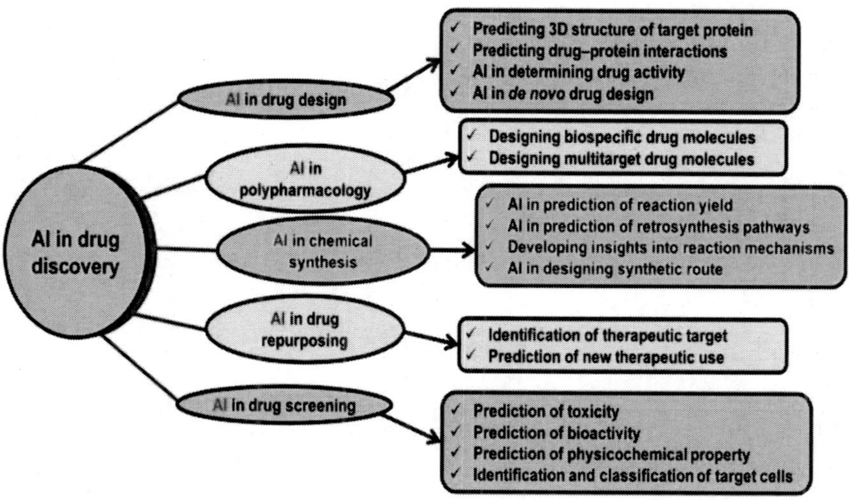

Figure 4. Application of AI in drug Analysis.

AI methods for designing the oligonucleotide therapeutics. This includes the small RNA interfering and CRISPR systems. A novel class is represented by the potential of a wide range of diseases. The AI techniques are therapeutics potential for a wide range of diseases. For example, eskip finder is a machine that skippers, which is a sequence of the identified effective predictions. The approaches that accelerate the development of the drugs. The optimized design and the reducing the time required for the validation.

Figure 4 explains the AI techniques and applications that are used in pharmaceutical analysis.

Drug Discovery by Stages

The various steps involved in the preclinical discovery drug stage are explained in the Figure 5.

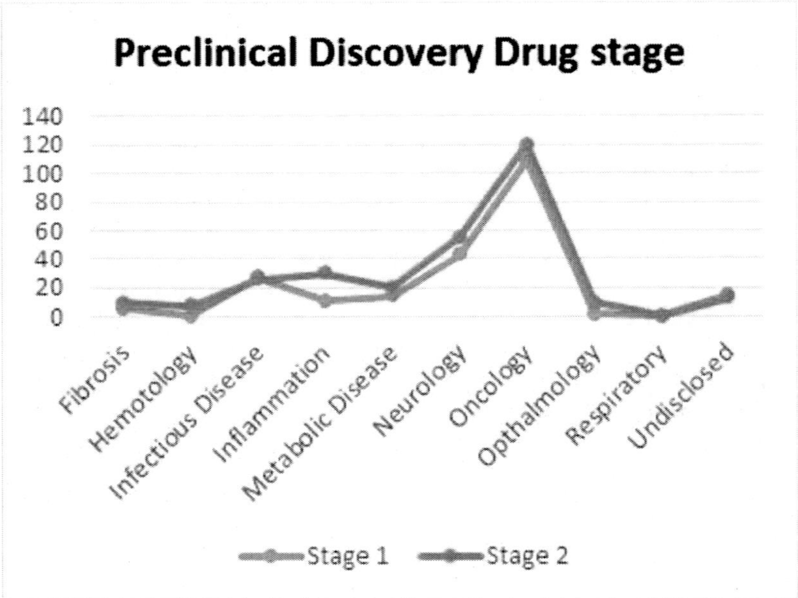

Figure 5. Different Stages of the Preclinical drug discovery.

Conclusion

AI has become an integral part of modern drug discovery, offering tools and techniques that significantly enhance the efficiency and effectiveness of the drug development process. By leveraging AI, researchers can explore vast chemical spaces, predict complex biological interactions, and design novel therapeutic agents with unprecedented speed and accuracy. However, the integration of AI into drug discovery is complicated. These include the need for high-quality data, the interpretability of AI models, and the ethical considerations surrounding the use of AI in healthcare.

References

[1] Bajorath, J. (2022). Artificial intelligence in interdisciplinary life science and drug discovery research. *Future Science OA, 8*(4), Art. no. FSO792.

[2] Bess, A., Berglind, F., Mukhopadhyay, S., Brylinski, M., Griggs, N., Cho, T., Galliano, C., &Wasan, K. M. (2022). Artificial intelligence for the discovery of novel antimicrobial agents for emerging infectious diseases. *Drug Discovery Today, 27*(4), 1099-1107.

[3] Chen, Z., Liu, X., Hogan, W., Shenkman, E., & Bian, J. (2021). Applications of artificial intelligence in drug development using real-world data. *Drug Discovery Today, 26*(5), 1256-1264.

[4] Dwivedi, K., Rajpal, A., Rajpal, S., Agarwal, M., Kumar, V., & Kumar, N. (2023). An explainable AI-driven biomarker discovery framework for non-small cell lung cancer classification. *Computers in Biology and Medicine, 153*, Art. no. 106544.

[5] Han, H., & Liu, X. (2021). The challenges of explainable AI in biomedical data science. *BMC Bioinformatics, 22*(S12), 1-3.

[6] Jiménez-Luna, J., Grisoni, F., & Schneider, G. (2020). Drug discovery with explainable artificial intelligence. *Nature Machine Intelligence, 2*(10), 573-584.

[7] Karger, E., &Kureljusic, M. (2022). Using artificial intelligence for drug discovery: A bibliometric study and future research agenda. *Pharmaceuticals, 15*(12), 1492.

[8] Kunduru, A. R. (2023). Machine learning in drug discovery: A comprehensive analysis of applications, challenges, and future directions. *International Journal of Orange Technology, 5*(8), 29-37.

[9] Rao, J., Zheng, S., Lu, Y., & Yang, Y. (2022). Quantitative evaluation of explainable graph neural networks for molecular property prediction. *Patterns, 3*(12), Art. no. 100628.

[10] Rajabi, E., &Kafaie, S. (2022). Knowledge graphs and explainable AI in healthcare. *Information, 13*(10), 459.

[11] Takagi, A., Kamada, M., Hamatani, E., Kojima, R., &Okuno, Y. (2022). GraphIX: Graph-based in silico XAI (explainable artificial intelligence) for drug repositioning from biopharmaceutical network. *arXiv:*2212.10788.

[12] Tripathi, A., Misra, K., Dhanuka, R., & Singh, J. P. (2023). Artificial intelligence in accelerating drug discovery and development. *Recent Patents on Biotechnology, 17*(1), 9-23.

[13] Askr, H., Elgeldawi, E., Ella, H. A., Elshaier, Y. A. M. M., Gomaa, M. M., &Hassanien, A. E. (2023). Deep learning in drug discovery: An integrative review and future challenges. *Artificial Intelligence Review, 56*(7), 5975-6037.

[14] Adadi, A., &Berrada, M. (2020). Explainable AI for healthcare: From black box to interpretable models. In *Embedded Systems and Artificial Intelligence* (pp. 327–337). Springer, Fez, Morocco.

[15] Ahmed, Z. U., Sun, K., Shelly, M., & Mu, L. (2021). Explainable artificial intelligence (XAI) for exploring spatial variability of lung and bronchus cancer (LBC) mortality rates in the contiguous USA. *Scientific Reports, 11*(1), 1–15.

[16] Aghaeipoor, F., Sabokrou, M., & Fernández, A. (2023). Fuzzy rule-based explainer systems for deep neural networks: From local explainability to global understanding. *IEEE Transactions on Fuzzy Systems, 31*(9), 3069–3080.

[17] Custode, L. L., &Iacca, G. (2023). Evolutionary learning of interpretable decision trees. *IEEE Access, 11*, 6169–6184.

[18] Dimitsaki, S., Gavriilidis, G. I., Dimitriadis, V. K., &Natsiavas, P. (2023). Benchmarking of machine learning classifiers on plasma proteomic for COVID-19 severity prediction through interpretable artificial intelligence. *Artificial Intelligence in Medicine, 137*, Art. no. 102490.

[19] Dhanorkar, S., Wolf, C. T., Qian, K., Xu, A., Popa, L., & Li, Y. (2021). Who needs to know what, when: Broadening the explainable AI (XAI) design space by looking at explanations across the AI lifecycle. In *Proceedings of the Designing Interactive Systems Conference* (pp. 1591–1602).

[20] Gerlach, J., Hoppe, P., Jagels, S., Licker, L., &Breitner, M. H. (2022). Decision support for efficient XAI services—A morphological analysis, business model archetypes, and a decision tree. *Electronic Markets, 32*(4), 2139–2158.

[21] Hailemariam, Y., Yazdinejad, A., Parizi, R. M., Srivastava, G., &Dehghantanha, A. (2020). An empirical evaluation of AI deep explainable tools. In *Proceedings of the IEEE Globecom Workshops (GC Wkshps)* (pp. 1–6).

[22] Hauser, K., Kurz, A., Haggenmüller, S., Maron, R. C., von Kalle, C., Utikal, J. S., ... &Sergon, M. (2022). Explainable artificial intelligence in skin cancer recognition: A systematic review. *European Journal of Cancer, 167*, 54–69.

[23] Hinnerichs, T., &Hoehndorf, R. (2021). DTI-voodoo: Machine learning over interaction networks and ontology-based background knowledge predicts drug-target interactions. *Bioinformatics, 37*(24), 4835–4843.

[24] Islam, M. S., Awal, M. A., Laboni, J. N., Pinki, F. T., Karmokar, S., Mumenin, K. M., Al-Ahmadi, S., Rahman, M. A., Hossain, M. S., &Mirjalili, S. (2022). HGSORF: Henry gas solubility optimization-based random forest for C-section prediction and XAI-based cause analysis. *Computers in Biology and Medicine, 147*, Art. no. 105671.

[25] Kobylinska, K., Orlowski, T., Adamek, M., &Biecek, P. (2022). Explainable machine learning for lung cancer screening models. *Applied Sciences, 12*(4), 1926.

[26] Langer, M., Oster, D., Speith, T., Hermanns, H., Kästner, L., Schmidt, E., Sesing, A., & Baum, K. (2021). What do we want from explainable artificial intelligence (XAI)—A stakeholder perspective on XAI and a conceptual model guiding interdisciplinary XAI research. *Artificial Intelligence, 296*, Art. no. 103473.

[27] Nguyen, T. L., Kavuri, S., Park, S. Y., & Lee, M. (2022). Attentive hierarchical ANFIS with interpretability for cancer diagnostic. *Expert Systems with Applications, 201*, Art. no. 117099.

[28] Ogiya, D., Chyra, Z., Verselis, S. J., O'Keefe, M., Cobb, J., Abiatari, I., Talluri, S., Sithara, A. A., Hideshima, T., Chu, M. P., Hájek, R., Dorfman, D. M., Pilarski, L. M., Anderson, K. C., &Adamia, S. (2023). Identification of disease-related aberrantly spliced transcripts in myeloma and strategies to target these alterations by RNA-based therapeutics. *Blood Cancer Journal, 13*(1), 23.

[29] Sharma, S., Rawal, Y. S., Pal, S., & Dani, R. (2022). Fairness, accountability, sustainability, transparency (FAST) of artificial intelligence in terms of hospitality industry. In *ICT Analysis and Applications* (pp. 495–504). Springer, Berlin, Germany.

[30] Streeb, D., Metz, Y., Schlegel, U., Schneider, B., El-Assady, M., Neth, H., Chen, M., &Keim, D. A. (2022). Task-based visual interactive modeling: Decision trees and rule-based classifiers. *IEEE Transactions on Visualization and Computer Graphics, 28*(9), 3307–3323.

Chapter 12

Data Preparation and Preprocessing

Richa Kahol[1,2,*]
and Atul Kathait[3]

[1]School of Biosciences, Apeejay Stya University, Gurugram, Haryana, India
[2]Medanta the Medicity Hospital, Gurugram, Haryana, India
[3]Assisstant Director-IQAC, Amity University, Noida , Uttar Pradesh, India

Abstract

Healthcare data compilation and conservation have gained importance. It has almost revolutionized medical research. Data management transformations in the healthcare sector are of prime importance to keep pace with the world because extraordinary technical advancements are occurring. The healthcare data management systems are currently centralized. Such system may experience system crashes in case of a high-scale medical emergencies. Decentralization of these systems may be a solution to this issue. Big Data Analytics (BDA) has immensely influenced the medical research and healthcare industry in recent years. Gradually, BDA has facilitated and improved routine duties in the healthcare sector. In this chapter, we introduce Big Data Analytics (BDA), its importance in the healthcare sector, decentralization of BDA in the healthcare sector, the challenges involved, and its advantages. We intend to discuss how decentralized BDA in the healthcare sector improves the quality of medical services. We propose a preliminary infrastructure for a decentralized healthcare ecosystem encompassing region, city, and country levels. Establishing such a system will ensure data security, promote the clinical data compilation and storage, and subsequently enable easy and secure sharing of patient data.

---

* Corresponding Author's Email: richabiotech03@gmail.com

In: Harnessing AI for Drug Discovery
Editor: Kolla Bhanu Prakash
ISBN: 979-8-89530-540-9

Introduction

Health has always been of prime importance and a susceptible topic for humanity. The public health index represents the quality of life, socio-economic standards, and political and cultural status of the society (Cunha et al, 2022). Therefore, to benefit civil society, it is essential to maintain and constantly improve the quality of its healthcare sector. Research-based prognosis is an integral part of medical research. Ample data is needed to have near-accurate research results. It is imperative to have processed data for better decision-making in terms of prognosis and treatment plans. Therefore, it is critical to provide accurate and precise data. Healthcare data is highly confidential and sensitive, so it must abide by specific rules and regulations to maintain its authenticity, integrity, privacy, and safety. In addition, decentralized BDA infrastructure in the healthcare sector guarantees a chronological order of data as well as its authenticity, privacy, and security (Angraal et al, 2017).

Big Data in Healthcare Sector

People view big data as an unorganized, highly complex, and heterogeneous collection of enormous data. Earlier, the enormity and convolutions of big data rendered it non-usable. It was almost impossible to collect and store this data. As computing technologies advanced, analysts stored, organized, mined, and analyzed big data to create valuable resources. Although there isn't a clear definition for "big data," De Mauro acceptably defined it as "the representation of information assets characterized by a high volume, velocity, and variety to require specific technology and analytical methods for its transformation into value" (Pablo et al., 2021; Berros et al., 2023). This statement indicates the "five Vs" of the big data, which are explained ahead, in the following sections (Figure 1).

Features of Big Data

Value

Collecting accessible big data is beneficial, but it remains worthless until it is transformed into valuable data (Pablo et al., 2021). It is imperative to fetch useful information from the massive volume of data. Usually, data generated from the healthcare service providers is enormous but it is neither organized nor scientifically passable. Contemplating and extracting scientifically constructive information from a large dataset makes raw data valuable. The primary aim of this process is to build on the current knowledge. Many healthcare institutes and public health research organizations invest huge capital to generate knowledge and value from the data repositories (Pablo et al., 2021; Berros et al., 2023).

Variety

In the healthcare sector, a variety of data is essential. There are various data sources and formats. Previously, data was mostly "unstructured" – for example, handwritten notes from nurses, prescriptions or hospital discharge summaries. But now, data may be structured, unstructured or semi-structure data. The sources to healthcare data are electronic health records (EHR), real-time data, images, audio-visual etc. Clinical trials, genetic research, electronic/ physiological or chemical sensors can produce data (Wang et al., 2015). Data is also gathered by emails, smartphones and wellness devices.

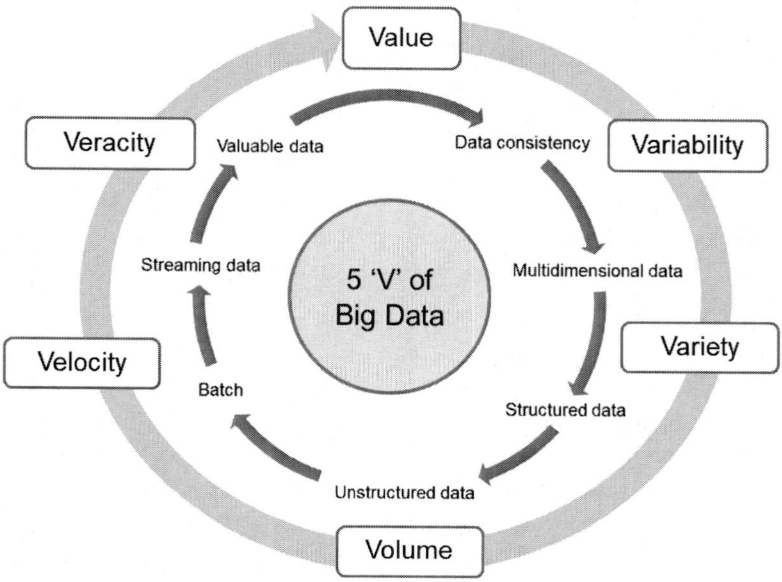

Figure 1. The five 'V' of big data.

Variability

It is defined as the consistency (or non-redundancy) of medical data over some time (Andreu et al., 2015).

Velocity

The speed at which data is generated, collected, transferred and processed defines the velocity. In the healthcare sector, the requirement of processed data varies according to the treatment pattern. In the case of diabetes treatment, blood glucose needs to recorded and stored for a long time to check the trends of fluctuating blood glucose level. Whereas in the case of heart surgery, blood glucose level and blood pressure needs to be optimized real time. On the other hand, the DNA sequencing data generated at the laboratory levels can be stored and processed for a very long time. Researchers recommend the use of big data analytics for the construction of predictive models. These models are efficiently used for early detection and response to medical emergencies. For instance, artificial intelligence (with machine language algorithms) can predict the outbreak and spread of an infectious epidemic (Alfred et al., 2021).

Veracity

The Veracity of the data defines the reliability, accuracy and quality of the data. Veracity highly depends on the source of data (Ishwarappa et al., 2015). In the healthcare sector, data is fetched from various sources and in the form of images, signals, or handwritten notes. Data accuracy is of prime importance in the medical industry and further treatment depends on the patient's medical history. Data validation and data accuracy are critical to avoid wrong diagnoses and mistreatment. Researchers suggest data mining, data cleaning, data validation, and data integration to maintain data consistency and preciseness (Berros et al., 2023).

Volume

As the name indicates, this attribute represents the enormity of the data in the healthcare sector. The data size ranges from 10^{12} bytes to 10^{21} bytes (Luo et al., 2016). The data in the healthcare sector is increasing exponentially. There are multiple sources of medical data generation. Also, the integration of new formats (sensory signals, three-dimensional images, and biological/omics) of data has added to the volume. Some studies (Zhou et al., 2022) recommend the use of natural processing language (NLP) to fetch valuable information from roughly written clinical notes (Berros et al, 2023).

What Is Big Data Analytics?

Big data analytics (BDA) involves mathematically or statistically analysing enormous raw data sets to uncover synchronous trends, relevant correlations, and suggestive data patterns; users can also use these analyses to develop predictive models based on their preferences. Conventional statistics such as clustering, correlation, and regression form the basis of this analysis (Hariri et al., 2019). The analysis technique are regularly re-evaluated and updated. These techniques have been modernized to keep pace with the exponentially increasing demand for BDA. Along with BDA techniques, analysts use several strategies to simplify BDA. For instance, these include:

- *Parallelization:* This strategy reduces the time to complete a massive task by splitting the task into smaller quantums and performing it simultaneously. The task is divided among numerous processors that conduct it simultaneously (instead of doing it sequentially). As a result the task is completed quickly (Wang et al., 2016).
- *The divide-and-conquer*: A slight difference distinguishes this strategy from "parallelization." Additionally, this method breaks the big problem down into smaller ones and resolves them all simultaneously. But, in the final step it combines the results to form one solution for the original (undivided) problem. Hence, the large problem is considered solved (Jordan et al., 2012).
- *Sampling:* With this strategy, the enormous amount of data is effectively reduced. It uses data sample criteria. Herein, selective data is organized, manipulated and analyzed to produce data trends and correlative models (Wang et al., 2012; Najafabadi et al., 2015).
- *Feature selection*: This is a traditional method of working with large datasets. In this method, a subset of large dataset is selected. The chosen subset includes pertinent highlights from the original data in a more accurate format. It is an effective for representing data mining and high-scale data (Guyon et al., 2002).
- *Instance selection*: This strategy is widely used as a data pre-processing method in many practical data mining tasks in machine learning and algorithms. It is possible to disable the run time and training sets during the classification or trial phases by using this method (Olvera et al., 2010).

- *Incremental learning*: This learning algorithm predominantly applies to streaming data, made of fresh data inputs. Incremental learning algorithms update according to the new data automatically each time (Wang et al., 2016).
- *Granular computing*: According to this strategy, dividing the larger task into smaller subsets (or granules) reduces the space they use. This approach has a highly impactful effect on reducing the uncertainty of objects in the search space. Currently, this strategy is under consideration for implementation in near future (Bargiela et al., 2016; Kacprzyk et al., 2016; Yager et al., 2018).

Classification of Big Data Analytics in Healthcare

There are many types of BDA. Data analytics vary according to the need of the content or data. Some of them are described here (Figure 2).

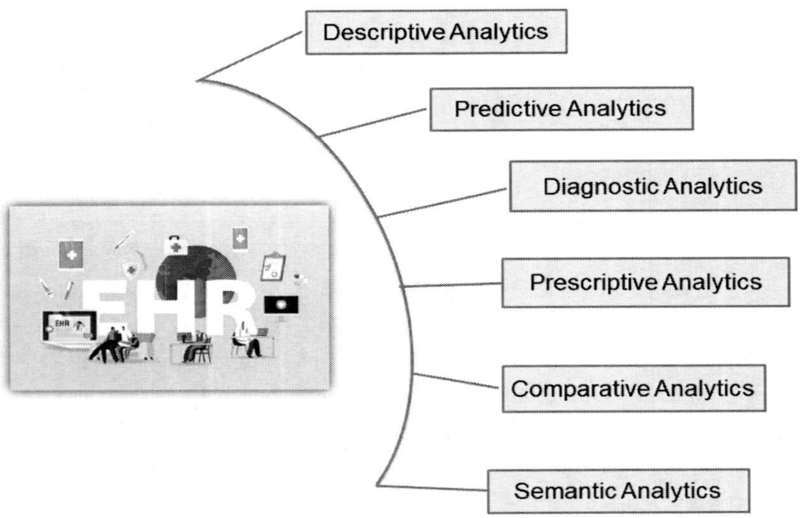

Figure 2. Types of big data analytics in the healthcare sector.

Descriptive Analytics

It is also known as unsupervised learning. This branch of healthcare data analytics makes use of past knowledge and data. It describes the current medical issue with references from the medical history. This analysis utilizes data mining, data storage, and healthcare business intelligence. It is a representation of non-analyzed raw data. Descriptive analytics merely states the effect of a parameter on the system without deep analysis. It collects, and categorically segregates data in an organized form. For instance, the average length of stay in the hospital, average discharge rate and recovery rate of the patients (Berros et al., 2023).

Predictive Analytics

As the name indicates, this branch of health care data analytics is involved in predicting the upcoming events (may or may not be complications) of the treatment a patient is undergoing. For example, the physician can reduce the dose of anti-glycaemic medicine based on the latest reports of the patient if the blood sugar levels have been under control for the last three months.

The Second example, the current ABG reports suggest the patient may develop post-surgery complications after a major heart surgery. Predictive analytics reveals possible outcomes by identifying and analysing the existing patterns in medical parameters. These predictive models have wide range of application. For instance, chemists can use the predictive model to determine how much drug stock they may need during an upcoming epidemic. Predictive models typically leverage huge data sets (unstructured data) to achieve the best results and maximum customer satisfaction in line with their business strategies. Machine-level languages are used to develop the predictive models (Berros et al., 2023). Predictive analytics involves developing data-driven models based on a coalition of multiple parameters. They understand and design the current trend from the present dataset and then use the learned information to prognosticate future outcomes or events. Thus, it is applied to forecast patient recovery or any uncertain occurrence. Also, the pharmaceutical industry uses it to prdict the success rate of any medicine (Pablo et al, 2022).

Diagnostic Analytics

Diagnostic analytics is one of the most essential and primary forms of healthcare data analytics. The purpose of diagnostic analysis is to identify the cause of a particular medical event. Every event in the human body appears to occur suddenly but may involve a cascade of events and biochemical changes. Diagnostic analysis detects the upcoming series of changes in the human body. Diagnostic analytics uses clustering and decision tree methods. It discovers the source of the problem and subsequently calculates its impact. Its results are based on extensive examination and precise analysis of the current datasets (Berros et al., 2023).

Prescriptive Analytics

Prescriptive analysis is the advised action plan that triggers the best possible outcome. To conduct a prescriptive analysis, a thorough and precise knowledge of the simulated subject is a prerequisite (Car et al., 2019). Prescriptive analysis takes into account all the prescribed drugs. Prescriptive analytics makes recommendations for a medical emergency, such as what should be done, how to achieve the best result, and what shall be the favourable outcome. The prescriptive analytics results are based on multivariate practical alternatives and studies. This multivariate analysis allows medical authorities to speculate on the upcoming issues or opportunities and recommend the best line of action at the moment while also considering the cause and long-term effects of the current decision (Berros et al., 2023). This analysis results in enormous data and cognizance of potential outcomes. This insightful branch of healthcare data analytics is still in the initial phases due to its elevated intricacy to make and authenticate alternatives for tangled ailments (Pablo et al., 2022).

Comparative Analytics

As the name indicates, the comparative analytics results are based on comparing two similar characteristic parameters. It develops a very concise probabilistic model by comparing similar entities or outcomes. This resultant model will provide insights into the relativity of the attributes. Their application is in the tracking trends of ordering pathological tests by the healthcare providers, readmission costs, or diagnostic testing impact on the healthcare practices (Car et al., 2019).

Semantic Analytics

This branch of healthcare analytics is an information-driven procedure that perceives and directs the specific semantic associations between the traits of the information (data) that results in valuable information. These results may address complicated and specific scientific queries (Mohammad et al., 2017). It is a magnificent method for answering queries related to biomedical, genomic, and pharmaceutical datasets. It recognizes the known obscure connections and produces details of new relationships between drugs' aftereffects, genotype-phenotype coordination and impact, pharmacological interaction analysis, and gene-drug interaction results. Semantic analytics utilizes an online stage to consolidate heterogeneous information with the help of linked-data methodology to expand the extent of scientific queries and deduce conceivable responses. The data concepts have formed the basis for the responses or answers, reflecting previous semantic relationships (Pablo et al., 2022).

Issues in BDA in Healthcare Sector

The main challenges related to big data in healthcare include those pertaining to data sources, data quality, data analytics, and legal matters. These are explained below (Figure 3).

Authentic Data Sources

- *Data repository*: These are the original store houses of the medical data. Data repository is a facility or a department within an institute where the patient data is stored for an extended period. These medical data storage houses spread across regions, cities, and countries. Free-flow data among these repositories can quickly modulate data sharing, but a lack of collaboration between the repositories hinders data sharing (Koutkias et al., 2019). Data repositories may be electronic health records, social media handles, and personal healthcare monitoring kits.
- *Clinical data*: It comprises all kinds of medical reports and observations. It includes the prescribed medicines by the doctor, the

diagnostic reports, the laboratory reports, and medical imaging results. Clinical reports are highly confidential documents.

- *Regular data update*: Periodical updating of old data is mandatory. The technical team of the institute is responsible for data management. It is important to update medical data on a daily basis, especially the number of mortalities. Accumulation of false data may affect the clinical research results (Berros et al., 2023).

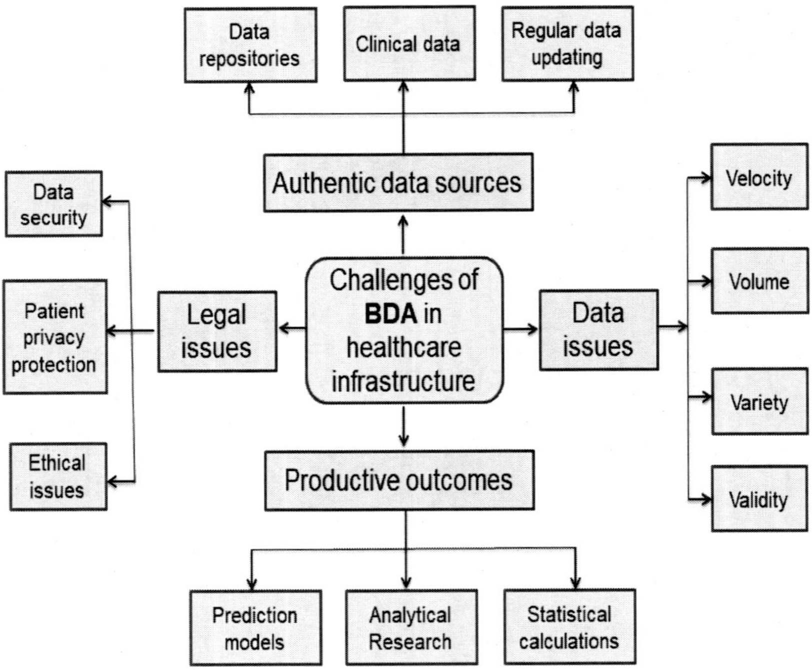

Figure 3. Challenges in Big Data Analytics in the Healthcare sector.

Data Quality Issues

Data quality alludes to every essential trait that explains big data analytics (as introduced in details in previous segment). In particular, there are four categories (4 verses):

- *Volume:* One must pay attention to the data size as it usually requires transforming a large amount of data. Also, one must consider the critical aspect of data scalability.
- *Velocity*: It is very important to maintain the speed at which data is collected, stored, and processed in the healthcare sector. It is important because the data influx is extremely high.
- *Validity:* Validating the data and its sources is a crucial step. Maintenance of accuracy and authenticity of the data demands clearance from a variety of sectors and sources.
- *Variety:* There are numerous sources of data. The implication of a variety of data sources (mobile phone applications) and maintaining an error-free inflow of it is challenging.

Productive Outcomes

- *Predictive models*: It represents the ability of big data analytics in healthcare to speculate and estimate the outcomes or any medical event. Continuous data monitoring is done to maximise the treatment's benefit, which is important.
- *Analytics applied*: Applying logical and scientific analytics is important in the medical sector because there are high possibilities of data misinterpretation, which can further lead to wrong treatment.
- *Statistical calculations:* statistical tests help convert raw data into valuable research results based on statistical formulae. It is important to apply correct statistical tests to the data. Professional biostatisticians perform these tests as it is a critical task.

Legal Issues

Several legal requirements must be met, and organizations must regularly update changes in the pertinent laws. Legal issues are equally important as data security issues. Therefore, equal time and effort is needed.

The legal issues in the field of health data include data security, maintaining patient privacy, and ethical issues. Health data contains highly confidential and personal information about patients. Therefore, it is the ethical responsibility of healthcare providers to ensure the security of the data.

Legal action may be required in case of a data security breach (Berros et al., 2023).

- *Data security*: Data security is a delicate matter that requires the highest level of sincerity. Medical facilities must ensure that licensed individuals have access to patient information. Encryption keys can also maintain confidential data in patient records.
- *Patient privacy*: To maintain patient privacy is of prime importance in the healthcare institutes and their research. Neither the academic nor non-academic study may disclose the patient's name. Patients are addressed as subjects and are identified by a unique number.
- *Ethical issues*: Institutional ethical committees ensure the unbiased approval of research studies. The committee includes representatives from various departments of the hospital, an external scientific officer, and a layman.

Requirement of Expertise

Another critical challenge of big data analytics in health care infrastructure is the requirement of experts. The healthcare sector is a very wide-spread industry. It includes the medical and paramedical professionals, technical, non-technical, academic, financial, clinical, and administrative staff. Therefore, it is imperative to stay abreast with technological advancements in all the associated sectors to maintain the ease of business and system functionality (Berros et al., 2023).

Decentralized Big Data in Healthcare Infrastructure

Decentralized architecture of BDA has emerged as a standard solution for many healthcare data-related issues. Initially, the goal was to solve data issues such as storing and sharing electronic health records (EHR) and electronic medical records (EMR). Artificial intelligence (AI) and blockchain methods have served as an asset to the healthcare data infrastructure. However, the decentralized BDA infrastructure consists of two components, mainly the open and closed end. The two ends of the e-healthcare data analytics are interconnected through data validators and depersonalization of data. This storage system is a distributed stockpiling of the prescribed medical data and health details of the patients. Enormous data is stored in an organised pattern. The 'closed' part of the system stores the personal information of the patient. It is a collection of EHRs. The doctor regularly updates the EHR and latest health status with electronic digital signatures (EDS). Thus, it improves data transparency (Kotsiuba et al., 2018) (Figure 4).

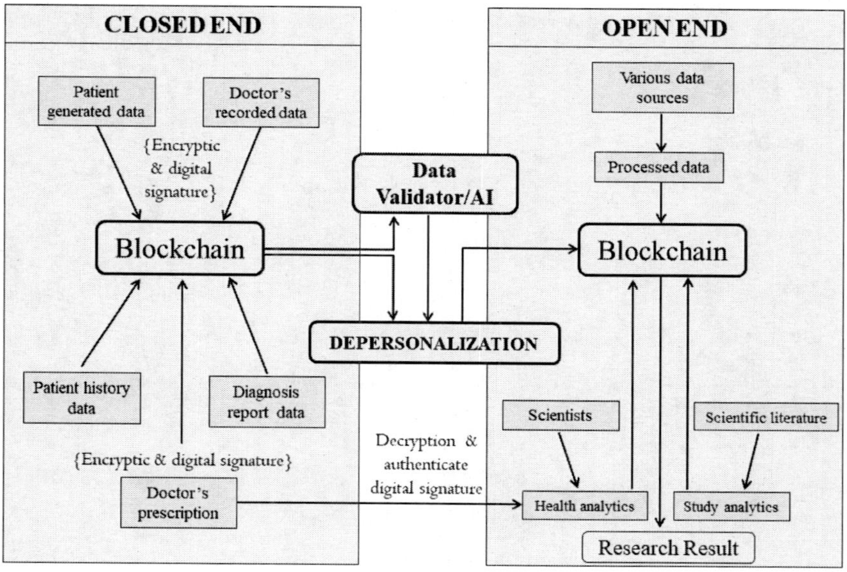

Figure 4. Layout for the architecture of e-healthcare data analytics decentralization.

Data validation is the first step of the system. Data is valid only if it bears an encrypted and digital signature (EDS) from the attending doctor. Data validation is followed by data depersonalization. Data depersonalization involves removal of unique patient identifiers. The data's confidentiality is preserved. After depersonalization, 'open' part of the system receives the data. Scientists can easily access data for research and analysis from such cloud storage systems.

Patient records or medical data sources are vast and are available in different formats (sensor signals, image informatics, genomics information, instrument readings, handwritten nursing notes, etc). When healthcare setups, such as multispecialty hospitals, clinical research units, or public medical research domains, apply data analysis strategies, they consolidate a wide range of types and formats of patient data. Big data analytics merges heterogeneous data, evaluates the data quality, develops a data model, validates the analysed data, and interprets the data results.

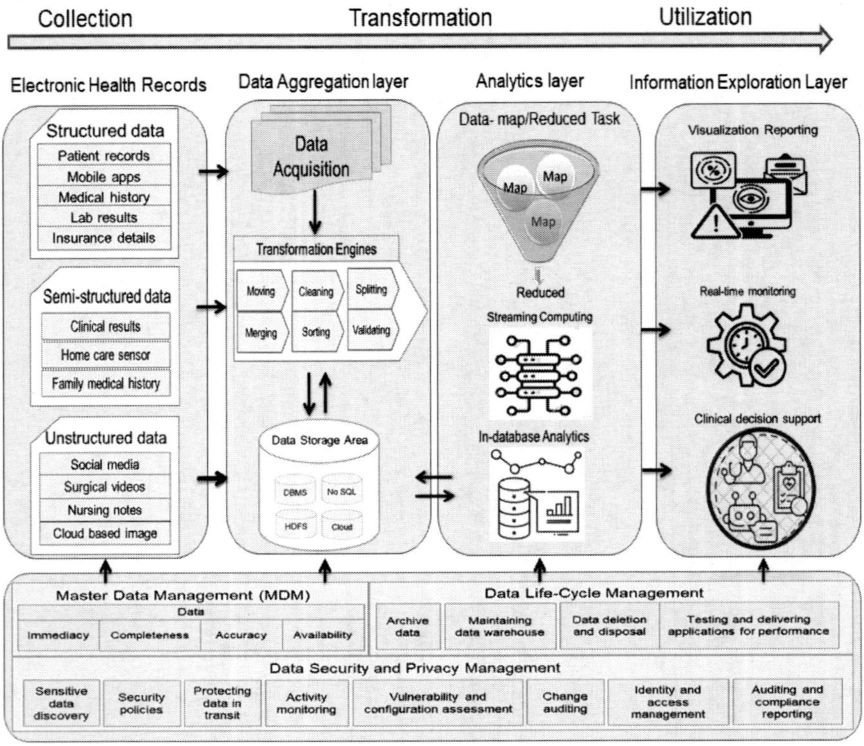

Figure 5. Workflow of the decentralized BDA infrastructure.

The raw patient data goes through three significant steps of procession. These steps are; ‘collection’, ‘transformation’ and ‘utilization’. ‘Collection’ comprises of collecting rough data from different data sources. ‘Transformation’ includes mining and refining of aggregated data. In this stage, the data is prepared for analysis. The final stage is ‘Utilization,” wherein the processed data is analysed and the research decisions are prepared (Figure 5).

The data inflow into the BDA infrastructure mainly consists of three units: the Master Data Management (MDM) system, the data lifecycle management system, and the data security and privacy system. This system has smaller sub-units. MDM comprises data immediacy, completeness, accuracy, and availability. A data lifecycle management system comprises archive data, data warehouse maintenance, data deletion and disposal, testing, and delivering performance applications. The data security and privacy management system includes data discovery capture, security policies, data protection in transit, vulnerability and configuration assessment, change reviews, identity and access management, audits, and compliance reporting (Figure 5).

The data inflow into the BDA infrastructure mainly consists of three units: the Master Data Management (MDM) system, the data lifecycle management system, and the data security and privacy system. This system has smaller sub-units. MDM comprises data immediacy, completeness, accuracy, and availability. A data lifecycle management system comprises archive data, data warehouse maintenance, data deletion and disposal, testing, and delivering performance applications. The data security and privacy management system includes data discovery capture, security policies, data protection in transit, vulnerability and configuration assessment, change reviews, identity and access management, audits, and compliance reporting (Figure 5).

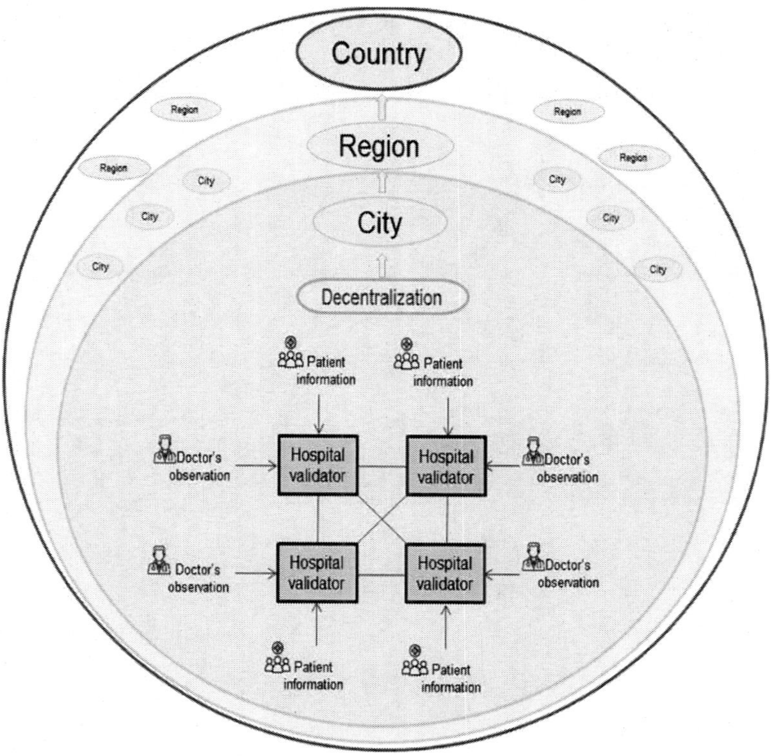

Figure 6. Decentralization of BDA infrastructure.

Decentralized BDA infrastructure is a sequential chain of connected blocks that form distributed hubs of healthcare data. Decentralized BDA ensures easy and secure access to medical data for healthcare personnel. Data is collected and compiled at the institute level and collectively at the 'city' level. The healthcare data from many cities are combined to orm a 'region.' Many regions combine to create a 'country' level. Thus, the data web continuously increases with each level. Figure 6 illustrates the diagrammatic representation of the decentralized BDA infrastructure (Figure 6).

The decentralized BDA infrastructure consists of "data auditors" and "data validators." Data authentication is necessary because data is collected from a variety of sources. Therefore, data auditors and data validators are required (Kotsiuba et al, 2019).

Advantages

The medical and other para-medical professionals highly support BDA for healthcare data. The benefits of data analytics in healthcare sector are very vast. Nonetheless, it can mostly be achieved in the following specific fields (Figure 7).

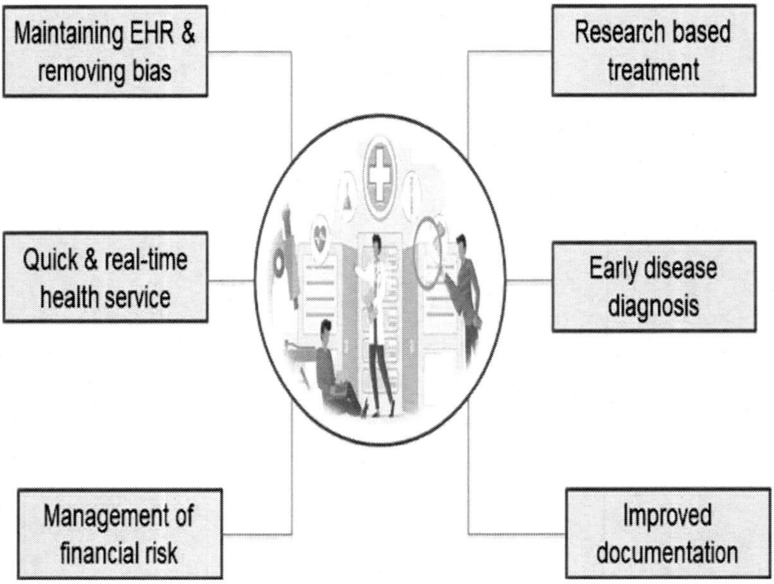

Figure 7. Advantages of Decentralized BDA in the healthcare sector.

Maintaining EHR and Removing Bias

It provides greater detail in the EHRs of individual patients. Decentralized BDA infrastructure ensures the preservation and maintenance of the EHR effectively. One master storage organization maintains the records, which helps improve non-biased research results. Since external researchers and scientists work on very large-scale data, there is no paucity of data or shortage of sample size.

Research-Based Treatment

In a decentralized BDA infrastructure, the research results and conclusions are available to a large audience on a large scale. These research results would be applicable to cases with specific treatments. Healthcare providers can refer to the case studies. These studies are very helpful in special medical issues. Clinical cohort studies correlate the treatment outcomes with the cause of diseases. Thus, these studies are scientifically valuable.

Quick and Real-Time Health Service

Deliver real-time alerts to healthcare providers by analysing health data at the collection point. Usually they are self-explanatory and user-friendly applications. With the help of these apps, patients may monitor their illness in real time.

On the other hand, medical professionals find it difficult to have timely access to the patient data whenever needed. Decentralized BDA ensures quick access to data and trackable patient records.

Early Disease Diagnosis

It promotes rapid and accurate diagnosis of high-risk diseases. Decentralized BDA can easily monitor the prevalence rate of any disease. It can identify the rate of symptoms, so medical professionals use it to diagnose diseases early.

Management of Financial Risk

It makes healthcare provisions more efficient and further reduces healthcare management costs. Precise treatment and early diagnosis also prevent patients from conducting unnecessary tests, thus reducing costs.

Improve Documentation

In addition to early diagnosis, patient database also helps in quick and hassle-free documentation of patient records. Furthermore, a well-structured data set is quite beneficial when doing research. This also saves the time and efforts needed in preparing documents required by higher authorities.

Data Analytics Advances Medical Research

The application of BDA in medical research has always been very promising. It improves public health organizations by early detection of disease outbreaks, timely prevention, and maintenance of satisfactory health safety. It also significantly contributes to sustaining a better quality of healthcare and thereby contributes towards expanding life span. BDA based research solutions provide high accuracy in prognosis and personalized medical treatment.

In addition to this, BDA applies to the hospital administration as well. It used to keep records of hospital discharges, readmissions, and latest ICU admissions. BDA is also used for staffing related challenges, such as recruitments, hiring, training, relocations, and labour cases.

Conclusion

Decentralized BDA infrastructure builds a healthcare data management ecosystem wherein medical data is easily accessible for the medical care providers, and at the same time it maintains utmost confidentiality. It has numerous benefits, such as maintaining patient records, real-time monitoring, and early diagnosis of disease. It also helps in improving the provision of medical services.

However, it is a strenuous topic, and its implementation is very demanding. There are many challenges related to this, from administrative to legal issues and from technical compatibility to social acceptance. There are many problematic concepts related to it. Nevertheless, the benefits are far more advantageous than investment. Not just hospitals, but small healthcare providers, public health organizations, pharmaceutical companies, and research institutes highly benefit from decentralized BDA infrastructure. It may lead to a transition from BDA to smart data analytics in medical data.

References

Alfred, R., Obit, J. H. The roles of machine learning methods in limiting the spread of deadly diseases: A systematic review. *Heliyon* **2021**, 7, e07371.

Andreu-Perez J., Poon C. C. Y., Merrifield R. D., Wong S. T. C., Yang G. Z. Big data for health. *IEEE J. Biomed Health Inf.* 2015;19:1193–208.

Angraal, S., H. M. Krumholz, and W. L. Schulz, "Blockchain technology: Applications in health care," *Circ. Cardiovasc. Qual. Outcomes*, vol. 10, no. 9, pp. 1–3, 2017.

Bargiela A., Pedrycz W. Fuzzy logic, systems, artificial neural networks, and learning systems, *Granular computing.* In: *Handbook on computational intelligence*, vol 1, p. 43–66. 2016.

Berros N., El Mendili F., Filaly Y., El Bouzekri El Idrissi Y. Enhancing Digital Health Services with Big Data Analytics. *Big Data and Cognitive Computing*. 2023; 7(2):64. https://doi.org/10.3390/bdcc7020064.

Car J., Sheikh A., Wicks P., Williams M. S. Beyond the hype of big data and artificial intelligence: building foundations for knowledge and wisdom. *BMC Med* 2019;17:143.

Cunha, João, Ricardo Duarte, Tiago Guimarães, César Quintas, Manuel Filipe Santos, Blockchain analytics in healthcare: An Overview, *Procedia Computer Science*, Volume 201, 2022, Pages 708-713, ISSN 1877-0509, https://doi.org/10.1016/j.procs.2022.03.095.

Guyon I., Weston J., Barnhill S., Vapnik V. Gene selection for cancer classification using support vector machines. *Mach Learn*. 2002;46(1–3):389–422.

Hariri, R. H., Fredericks, E. M. & Bowers, K. M. Uncertainty in big data analytics: survey, opportunities, and challenges. *J Big Data* **6**, 44 (2019). https://doi.org/10.1186/s40537-019-0206-3.

Ishwarappa K., Anuradha J. A big introduction on big data 5 V's characteristics and Hadoop. *Procedia Comput Sci*. 2015;45:319–24.

Jordan M. I. Divide-and-conquer and statistical inference for big data. In: *Proceedings of the 18th ACM SIGKDD international conference on knowledge discovery and data mining.* New York: ACM; p. 4. 2012.

Kacprzyk J, Filev D, Beliakov G. (Eds.). *Granular, Soft and fuzzy approaches for intelligent systems: dedicated to Professor Ronald R. Yager* (Vol. 344). Berlin: Springer; 2016.

Kotsiuba, I., A. Velvkzhanin, Y. Yanovich, I. S. Bandurova, Y. Dyachenko and V. Zhygulin, "Decentralized e-Health Architecture for Boosting Healthcare Analytics," *2018 Second World Conference on Smart Trends in Systems, Security and Sustainability (WorldS4)*, London, UK, 2018, pp. 113-118, doi: 10.1109/WorldS4.2018.8611621.

Koutkias, V. From Data Silos to Standardized, Linked, and FAIR Data for Pharmacovigilance: Current Advances and Challenges with Observational Healthcare Data. *Drug Saf.* **2019**, 42, 583–586.

Luo J., Wu M., Gopukumar D., Zhao Y. Big data application in biomedical research and health care: a literature review. *Biomed Inf Insigh*ts 2016;8:1−10.

Mohammadhassanzadeh H, Van Woensel W, Abidi SR, Abidi SSR. Semantics-based plausible reasoning to extend the knowledge coverage of medical knowledge bases for improved clinical decision support. *BioData Min.* 2017 Feb 10;10:7. doi: 10.1186/s13040-017-0123-y. PMID: 28203277; PMCID: PMC5303296.

Najafabadi M. M., Villanustre F., Khoshgoftaar T. M., Seliya N., Wald R., Muharemagic E. Deep learning applications and challenges in big data analytics. *J Big Data.* 2015;2(1):1.

Olvera-López J. A., Carrasco-Ochoa J. A., Martínez-Trinidad J. F., Kittler J. A review of instance selection methods. *Artif Intell Rev.* 2010;34(2):133–43.

Pablo R. J., Roberto D. P., Victor S. U., Isabel G. R., Paul C., Elizabeth O. R. Big data in the healthcare system: a synergy with artificial intelligence and blockchain technology. *J Integr Bioinform.* 2021 Aug 18;19(1):20200035. doi: 10.1515/jib-2020-0035. PMID: 34412176; PMCID: PMC9135137.

Wang L., Alexander C. A. Big data in medical applications and healthcare. *Am Med J* 2015;6:1−8.

Wang X. Z., Dong L. C., Yan J. H. Maximum ambiguity-based sample selection in fuzzy decision tree induction. *IEEE Trans Knowl Data Eng.* 2012;24(8):1491–505.

Wang X., He Y. Learning from uncertainty for big data: future analytical challenges and strategies. *IEEE Systems, Man, & Cybernetics Magazine* 2016.

Yager R. R. Decision making under measure-based granular uncertainty. *Granular Comput.* 1–9. 2018.

Zhou, H. Developing Natural Language Processing to Extract Complementary and Integrative Health Information from Electronic Health Record Data. *In Proceedings of the 2022 IEEE 10th International Conference on Healthcare Informatics (ICHI)*, Rochester, MN, USA, 11–14 June 2022; pp. 474–475.

Chapter 13

AI-Driven Drug Discovery: Transforming the Pharmaceutical Landscape

Pooja Dehankar*
and Susanta Das
[1]Ajeenkya DY Patil School of Engineering, Affiliated to Savitribai Phule Pune University, Charoli Bk. via Lohegaon, Pune, Maharashtra, India
[2]School of Engineering, Ajeenkya DY Patil University
Charoli Bk. via Lohegaon, Pune, Maharashtra, India

Abstract

In the traditional drug development processes obstacles, such as time, cost, and inefficiency are being encountered by the growing complexity of diseases accompanied by the need for new medicines. The advent of digital technology has seen machine learning (ML) algorithms come into play in supporting basic pharmaceutical processes like novel medication discovery. Rapid adoption of Artificial Intelligence (AI) techniques is forcing a paradigm shift in pharmaceutical research. This chapter envisages improved clinical trials, expedited drug development pipelines personalized therapy based on predictive models, and changes to pharmaceutical repurposing approaches. It outlines the old way of researching drugs and associated issues that make it too protracted a process. It considers how advancements in AI for pharmacy may impact the future creation of medications. However, this chapter discusses how ML can revolutionize pharmacy R&D and talks about some challenges related to data quality, model interpretability, ethics, and international cooperation.

---

* Corresponding Author's Email: poojadehankar@dypic.in

In: Harnessing AI for Drug Discovery
Editor: Kolla Bhanu Prakash
ISBN: 979-8-89530-540-9

Keywords: drug development processes, machine learning, pharmaceutical repurposing, Artificial Intelligence, model interpretability

Introduction

Conventional research requires several stages to identify new drugs and get them authorized for sale. Shortening the duration needed to discover a novel drug is crucial. Drug repurposing is becoming increasingly significant as a means of finding new applications for medicines presently on the market. Repurposing is usually achieved by deliberate techniques (incidental observations). It talks about a few strategies for coming up with novel uses for FDA-approved drugs. As a result, repurposing drugs has emerged as a successful method for developing new drugs, despite its difficulties. It offers a fresh perspective on repurposing existing medications for new applications (Parvathaneni et al., 2019).

Revolutionary progress is expected when artificial intelligence (AI) is incorporated into a contemporary, evidence-based drug discovery platform. While contemporary medicine keeps playing a crucial part in the fight against numerous serious illnesses, effective treatments for other chronic ailments are still pending discovery. Furthermore, a lot of contemporary medications have side effects and usually need to be replaced with safer, more recent versions. Therefore, traditional remedies made from plants may serve as a different kind of knowledge base for the study of drug discovery. Nevertheless, the methods used to investigate this strategy are no longer appropriate (Khan et al., 2021).

Scientists working on drug development often must give up one of their ideas to support someone else's, and they must tailor their scientific efforts to the demands of the project to which they contribute. In contrast, scientists in the university setting normally pursue their ideas and interests, which are often sparked by the findings of their earlier work or sporadic scientific "hot topics." Interaction between the scientific and drug discovery domains is crucial, though. Chemistry and other life sciences are vital to the process of discovering new drugs because they advance our understanding of how diseases work, which in turn leads to the development of pharmacological and biological remedies. The current phase of knowledge generation in the life sciences is exponential, and it takes place in the university settings (Herrling, 2005).

In the past, new drug entities were primarily derived from natural materials for drug discovery. Eventually, though, research based on combinatorial chemistry and high-throughput synthesis assumed primacy. New technologies like artificial intelligence and cutting down on the expense and duration of early drug discovery, ultra-high-throughput drug screening are being used extensively, however, they have not changed much yet (Berdigaliyev & Aljofan, 2020).

The development of novel pharmaceuticals aims to reduce side effects, improve quality of life, save costs to the healthcare system, and significantly extend patient life in addition to treating or preventing diseases for which there is no therapeutic alternative. The scientific, technological, and regulatory obstacles involved in comprehending the molecular mechanisms of action of drugs for complicated diseases are the source of these onerous costs and time constraints.

Investment in extremely complex technology, cutting-edge production techniques, and innovative research methods to address the process's ever-increasing cost and duration are thus necessary to achieve real success (Al Qaraghuli et al., 2017).

The development of an assumption that links the activation or inhibition of a protein or signaling path to a healing endpoint typically begins with academic research (Hughes et al., 2011). A fundamental study is conducted to find a target and confirm the choice. Target selection and validation are followed by a transition to a lead discovery phase to support a specific drug development attempt. The lead discovery stages necessitate thorough and thorough searches to locate a biological candidate or small molecule with drug-like properties that can advance into preclinical, and clinical trials, and, if successful, approval (Al Qaraghuli et al., 2017).

The pharmaceutical industry has experienced unprecedented and indisputable success in recent times. Though commendable, the medication process of development and discovery is among the most complex human activities because it must strike a balance between safety and efficacy in terms of health benefits at a suitable therapeutic index. For patients, this procedure frequently means the difference between life and death; scientists and doctors who discover, create, and deliver pharmaceuticals for illness prevention, treatment, and other diseases are the ones who will determine whether or not their patients survive this process. Modern medicine, which began with the discovery of aspirin at the close of the 1800s, emerged from organic synthesis and has profoundly altered both our way of life and death (Nicolaou, 2013; Nicolaou & Montagnon, 2008).

The complicated architecture of the biomolecules that cause diseases like AIDS, cancer, autism, and Alzheimer make it difficult for medical researchers to find treatments for these conditions. Considering recent changes in several elements, such as dietary habits, environmental conditions, and migration patterns in human existence. It is becoming important to create new antidrugs that are both effective and free of negative effects. Cheminformatics addresses drug discovery through contemporary drug discovery methodologies, hence resolving intricate problems in the conventional drug discovery system. Chemical informatics instruments aid medicinal chemists in comprehending intricate molecular structures. Finding Novel Chemical Entities (NCE) is the main objective of the rapidly evolving multidisciplinary field of cheminformatics, as it leads to the creation of new molecules. Additionally, it is essential for gathering, preserving, and assessing chemical data (Begam & Kumar, 2012).

Drug delivery technology has advanced dramatically as a result of the necessity to handle issues that are getting bigger every day. Three essential physicochemical properties—solubility, stability, and bioavailability—were the focus of the first generation of medicines. Cell membranes and the blood-brain barrier are examples of biological barriers that have grown to be significant roadblocks. As a result, second-generation drug delivery methods that are specially made to get past these biological obstacles have been developed. These improvements made use of techniques including improved penetration, tailored distribution to particular tissues or cells, and defense against the body's defense mechanisms being undermined. The ultimate goals of these second-generation strategies were to reduce adverse effects and increase therapeutic efficacy (Pankevich et al., 2014).

It became apparent that even with the advances in second-generation drug delivery technology, a comprehensive approach was needed to address both physicochemical and biological restrictions. As a result, third-generation medication delivery systems were created. Third-generation drug delivery systems are designed to address problems related to biology and physiochemistry. Along with thinking of ways to get over biological barriers and improve drug delivery to the target site, they aim to maximize the qualities of the medication, such as its solubility, stability, and release kinetics (Yun et al., 2015).

Much effort has gone into addressing low water solubility by changing excipients or other formulation methods. Another key focus has been on creating targeted and regulated delivery systems for complicated medications such as peptides and proteins. To improve therapeutic profiles, the final stages of drug development increasingly make use of molecular biology, genomics, recombinant technologies, and monoclonal antibodies. However, there are still some significant obstacles that need to be achieved to successfully overcome drug intolerance concerns, human physiology limitations, and pathogenic resistance issues. These require an acceleration of new ways of thinking for potential new solutions for improved outcomes (Arden et al., 2021).

AI for Smarter Drug Development

The healthcare business is embracing AI-based techniques, which impact areas including drug creation, diagnosis, prevention, and clinical recommendation making. These techniques are clever, versatile, and low-cost. Data science techniques can expedite the long and expensive process of creating new medications by being used in target identification, De novo molecular design, drug repurposing, retrosynthesis, and the prediction of reactivity and bioactivity, FDA approval, and post-market study. When estimating the characteristics of tiny molecules, deep neural networks (DNNs) can increase prediction power, and one-shot learning can be employed if a substantial quantity of trial data is unavailable. Considering labeling constraints, biological variability, and technical and human errors related to the underlying data is essential for developing predictive models that work. It is challenging to depict the experimental data quantitatively or with computer assistance. AI is currently being used to generate trial representations that facilitate data classification and, eventually, the creation of predictive models. Recently, artificial intelligence (AI) has begun to be applied more widely throughout society, with the pharmaceutical sector leading the way in benefits. AI has numerous important practices in the pharmaceutical sector, including clinical trials, drug repurposing, drug research and discovery, and productivity enhancement. Goals are accomplished more quickly and human labor is reduced as a result. There is also debate over tools and techniques for enforcing AI, the issues that remain unresolved, and the future of AI in the pharmaceutical sector (Kempt & Nagel, 2022).

The development of medications may be viewed as the pinnacle of pertinent translational scientific endeavors that increase human happiness and resilience. Still, the method of creating a unique medication is costly, slow, and complex. The pharmaceutical sector is currently engaged in a rigorous and stimulating brainstorming session to identify strategies for cutting expenses and accelerate the development of new drugs. AI and its components are now more accessible across all domains thanks to the use of classified huge data, significantly improved computing power, and cloud storage. AI has brought new life to computer-aided drug research. Technology breakthroughs in computer hardware and software that collaborate with a variety of problem aspects and the unrestricted promotion of ML across a wide spectrum of scientific disciplines support this progress. Computer-aided drug discovery has made extensive use of ML algorithms. Due to their capacity to generate nonlinear input-output pertinencies and enable automatic attribute elicitations from the input data, artificial neural networks (ANNs), which are DL techniques that employ numerous buried processing layers, have gained increased attention recently. These properties of the DL methodology supplement conventional ML methods that use human-generated molecular descriptors. Medicinal chemistry has advanced because of a large amount of original pessimism surrounding the use of AI in pharmaceutical discovery, beginning to vanish. It is predicted that AI will accelerate the search for new and improved medications in a timely, cost-effective, and smart way when combined with contemporary experimental technical knowledge. The use of DL-facilitated techniques has just recently begun to address several crucial drug development concerns. The examination of many of the most challenging and intriguing problems will surely benefit from several industrial innovations, for example, "message-passing paradigms," "spatial-symmetry-preserving networks," "hybrid de novo design," and additional ingenious ML exemplars. Drug development will improve significantly with the help of AI-powered open data dissemination and model enhancement (Sarkar et al., 2026).

The most important phase in the drug research and discovery process is drug screening or the assessment of potential therapeutic compounds' biological activity. Due to advancements in synthesis technologies, including automated synthesis and combinatorial chemistry, in addition to the discovery of an increasing number of new pharmacological targets, high-throughput screening assays have become essential for timely stage drug discovery in recent years.

Because they provide quick and sensitive analysis detection and work with tiny sample volumes, bioluminescence, and chemiluminescence are appropriate detection methods for high-throughput screening. Applications for AI analysis based on networks for drug development are repurposing and drug screening. Before choosing a candidate medication to move forward with clinical trials, prospective compounds are found and optimized through the process of drug screening (Roda et al., 2003).

Biological experimentation is a difficult, costly, and time-consuming method of drug screening. Therefore, the applications for network-based analysis of biology provide a viable substitute. To perform drug screening, it is necessary to identify drug-target interactions (DTIs). Some of these innovative DTIs can be used to identify new anticancer drugs with known targets (Hinnerichs & Hoehndorf, 2021).

The drug development and discovery pipeline is a hard, multifaceted and depends on many factors. The ML techniques offer a range of tools that can enhance discovery and decision-making if copious amounts of high-quality data and well-articulated queries are combined together. Each stage in the drug discovery process can benefit from the use of ML. These include target validation, digital pathology clinical trials data processing, and predictive biomarker identification. The methodologies and contexts within which they have been applied have varied; some approaches yielded penetrating insights as well as accurate predictions. One major challenge is that machine learning (ML) often produces findings that cannot be interpreted or repeated thus limiting its usefulness. In all domains there still needs to be produced high-dimensional data which is comprehensive and systematic. By taking additional steps to address these challenges and increase awareness about what is required for evaluating machine learning models, the adoption of ML can enable data-driven decisions thereby helping expedite the delivery of fewer failed drug discovery projects (Vamathevan et al., 2019).

AI innovation is a top priority in drug discovery through the assembly of pharmacological data and ML technique development. Artificial Intelligence does not depend on theoretical advancements; its value lies in converting medical data into research by reusable techniques. Many approaches are generally taken into consideration when discussing ML algorithms (Lavecchia & Di Giovanni, 2013).

As simple features are converted into complex features through multi-layer feature extraction techniques, through the use of currently available initialization data, deep learning techniques will automatically accelerate certain features. In conventional ML models, a large number of designed characteristics are executed by hand. One benefit of using deep learning techniques is that more accurate results can be obtained because there are fewer generalization errors. Deep learning techniques that are said to be unique. Le Cun et al. (2015), Angermueller et al. (2016), and Schmidhuber (2015) have provided an overview of deep learning algorithms, while Goodfellow et al. (2016) have provided extensive information on deep learning methods in the Deep Learning writings. Various AI computations are used in data analysis and prediction during the process of finding and developing drugs.

The ability of ML to swiftly and precisely analyze big, complicated datasets is one of its main benefits throughout the process of finding new drugs. Computational techniques for drug improvement have been available for a while, but their applicability hasn't always been trusted. Nonetheless, the discipline of chemoinformatic now has enough data due to the expanding availability of publicly accessible pharmacological databases to enable the development of ML techniques that can accurately predict drug activity (Hopkins, 2009).

One of the most important components of the drug development process is predicting the interactions that tiny compounds will have with proteins. Accurate predictive models are particularly being utilized more and more to screen for side effects or preselect possible lead compounds from big chemical databases. Traditional silico methods focus on predicting interactions with a single target, while new chemogenomic approaches use a cross-target viewpoint. ML algorithms can recognize patterns and relationships in big amounts of data that may not be immediately visible through conventional analysis techniques (Jacob & Vert, 2007).

Researchers can find possible medication candidates that were missed by more conventional screening techniques. Large datasets of chemical and biological data can be used to train ML algorithms to predict the properties of new substances. A new technique called "in silico target fishing" uses data from increasingly accessible biologically annotated chemical databases to anticipate a compound's biological target determined solely by its chemical composition (Jenkins et al., 2006).

Developing a novel naturally active chemical depending on already produced compounds and quantitative arrangement and activity connections that have been established are among the main issues with drug discovery. Afterward, new medication proposals with improved biological activity and a higher selectivity profile for a particular therapeutic target are made using this knowledge. A vast amount of data has been generated and collected due to improvements in the speed and capacity of computers. Finding meaningful and intelligible patterns from these enormous in silico libraries is therefore a formidable task. Because traditional statistical methods are insufficient, the development of data mining tools for drug discovery is the first step toward establishing data mining as a significant research direction (Agatonovic-Kustrin & Morton, 2016).

Data mining is an automated process that extracts useful information, sometimes unknown, from huge databases or data sets by employing sophisticated search strategies and algorithms to find patterns and connections in pre-existing large scale databases. Data mining can be used to build a framework for linking a collection of molecular descriptors to biologically significant features, which will help in understanding the structure-activity relations (SARs) of novel compounds. Using this model, one can prioritize new compounds for additional screening and forecast important property values of those compounds. One major focus area within the chemoinformatics spectrum for data-mining methodologies is the process of searching through huge databases using computer technologies for novel ideas that have a greater chance of having a significant binding affinity to the target protein (Lavecchia, 2015).

The main objectives are to speed up drug discovery, improve the safety profiles of already-approved medications, offer cutting-edge treatment modalities, and strengthen and comply with manufacturing processes. Although they stress the necessity for cautious application, recent FDA publications provide an optimistic view of these instruments' potential. This has increased demand for retraining workers using these technologies and made it possible for cutting-edge applications in cutting-edge medicines including CAR-T cells, mRNA-based therapies, gene editing, CRISPR-Cas9, and customized medicine (Niazi, 2023).

Determining Drug Targets in the AI/ML Era

Before an experiment on a drug target, the available data can be used by AI/ML systems to learn from it and predict what it might do (Schneider, 2018). High-throughput screening has been greatly enhanced by AI, allowing quicker and more accurate analysis of vast numbers of chemicals (Chen et al., 2018). Combining various data—featuring clinical and genomic information with proteomics information—led by AI/ML has ushered in improved knowledge about the biology of the disease and potential targets (Mamoshina et al., 2016). Proteins' three-dimensional structures, and the potential interactions between them and other compounds, with the aid of AI/ML approaches can bring about betterment in the creation of drugs (Wallach & Heifets, 2018). ML can personalize drug therapy by considering patient-specific information leading to more personalized and effective treatment - this is according to (Topol, 2019). AI-driven platforms enhance inter-institutional collaboration by pooling together varied data and knowledge which may lead to more innovative ways of identifying pharmacological targets (Alvarez-Uria, 2014). It accomplishes the task through a technique known as repositioning or repurposing drugs; finding new potential targets for already approved medications increases their therapeutic value (Dudley et al., 2011). This is how ML reduces the total cost of the drug development process by enhancing speed and accuracy during target discovery - according to (Mak & Pichika, 2019), who also stated that. Biological questions help define constraints on acceptable solutions as part of machine learning algorithms applied at large scale across complex biological networks to arrive at underlying mechanisms as well as potential therapeutic targets (Zitnik et al., 2019). To identify safer drug targets, ML models are applied to estimate the toxicological properties of chemicals (Zhang et al., 2017). AI and ML simplify the design and management of clinical trials, making it possible to forecast patient responses to specific medicines (Collins & Varmus, 2015). Drug targets can be found more accurately when artificial intelligence (AI) is used to analyze real-world information obtained from sources such as electronic health records (Sherman et al., 2016). AI/ML methods also support the identification of biomarkers related to diseases, which may be critical for targeted therapy (Camacho et al., 2018). AI can aid in developing medication combinations that may be more effective adjuvants for treatment-resistant diseases (Preuer et al, 2018.)

Drug Discovery

Drug discovery is a process of detecting compounds of use in the remedy and treatment of sickness. The procedure involves- candidate identification, screening, testing, characterization, and validation optimization as well as final synthesis for therapeutic efficacy. Their success in basic research will need to carry into the start of medication development before clinical trials are possible. The new drug needs to go through multiple stages of the drug development process to deliver a medicine that is compliant with all regulatory requirements and safe as well as effective. Turning a scientific discovery into something that can be marketed as medication takes many years and thousands of hours of labor. Out of every million molecules found, only one is studied in sophisticated clinical trials before being made available to patients. Among the phases of drug development and discovery are: Target identification, Validating the target, Identifying and optimizing leads, and Characterizing the product. Development and conceptualization, Preclinical research, clinical experimentation, novel drug submission acceptance, and new drug investigation (Deore et al., 2019).

The main objective of pharmaceutical companies is to create and bring new medications to market through the difficult processes of research and development. New drugs are continually needed by healthcare systems to treat unmet medical requirements in a range of therapeutic areas. Finding a candidate for additional development is the last step in the discovery process, which also includes hit identification, lead creation, validation, target identification, and optimization. On the other hand, development includes improving chemical production and formulation, studying animal toxicity, conducting clinical studies, and finally receiving regulatory approval. These two labor-intensive and costly procedures come at a time when the sector is facing pressure from increasingly strict regulations, environmental concerns, and declining revenue from patents running out. Since these problems have recently hurt R&D productivity, industry, academia, and government research institutions must work together more to develop novel solutions and maintain a shared goal of producing high-quality medications (Sinha & Vohora, 2018).

Precision, dependability, and expandability in protein-ligand binding affinity forecasts could significantly speed up the drug development process. The following problems persist despite great efforts: generalizations for out-of-distribution (OOD) for poorly defined proteins or compounds from chemical scaffolds or unlabeled protein families; scalability to billions of compounds; and uncertainty measurement of individual predictions. The DL framework based on sequences relying on affinity tackles the above-identified problems. Unique uncertainty regularized optimization, residue-estimation-based effective uncertainty quantification, and A model of protein language informed by structure are all produced by TrustAffinity. State-of-the-art computational techniques are greatly outperformed by TrustAffinity (Badkul et al., 2024).

Personalized Treatment Driven by Predictive Models

The high rates of clinical attrition make the growth of popular medications costly and long. In the context of customized medicine, this is partially due to the break in the drug's translatability from the bench to the clinic. By utilizing data that is widely available throughout all stages of the process, AI-driven platforms combined with automatic modeling have proven crucial in speeding up the drug improvement process. AI has the potential to overcome shortcomings and uncertainties that arise in the traditional drug development procedure, minimize the need for human involvement, and close the translational hole in understanding the relationships between medications and illnesses (Chakravarty et al., 2021).

The technique of inferring implicit information about a person from the outcomes of a reference group of "similar" patients undergoing different treatments is known as "reference class forecasting," and it occurs when clinical trial data is used to support decisions for specific patients. The wide reference class of patients who meet trial eligibility requirements has typically been highlighted in evidence-based medicine. However, the differences among trial participants (as well as those in actual practice) can have an impact on the desired outcome and potential benefits. Personalized medicine, in all its manifestations, aims to produce more patient-specific effect estimates by narrowing the reference class, hence facilitating more customized therapeutic decision-making (Kent et al., 2018).

Shift in Pharmaceutical Repurposing Strategies

Repurposing drugs for new uses is a time- and money-efficient approach with high success rates that can dramatically reduce the risk involved in developing treatments for uncommon diseases. Given the high failure rates, high costs, and the sluggish pace of new medical discovery, repurposing "conventional" medications to treat uncommon and joint disorders is starting to seem like a viable option. Delisted medications, lower research costs, and quicker development times are to blame for this. Technical and legal limitations must also be taken into account. Repurposed drug competitors have been formed through strategic decisions based on data analysis, interpretation, and exploratory methodologies. Innovative novel medical uses are produced when computational and experimental approaches are combined. It is a novel approach of using drugs that are safe for humans again to treat rare and challenging-to-treat illnesses. It is an extremely successful process for finding and developing new drugs. Repurposing outdated drugs has allowed a number of pharmaceutical companies to create innovative treatments. Repurposing medications is a fast, efficient, and practical option with generally less risk than creating a brand-new medication from scratch (Mishra et al., 2024).

There is growing agreement that Drug Repurposing, Repositioning, and Rescue (DRPx) has a significant and long-lasting impact on the pharmaceutical sector. The efforts of pharmaceutical corporations and DRPx-focused businesses, along with their activity and productivity, seem to give some optimism in finding solutions to the various issues facing the industry right now. Only until the dynamic variables of successful and innovative business models are recognized and comprehended can these efforts be maintained and increased (Naylor et al., 2015).

Conclusion

Early-stage drug development can now be accelerated by machine learning, which combines the phases of molecule production and filtering into a single architecture that takes advantage of medicinal chemists' design preferences and experience. However, because of the large search area, creating machine learning models that can satisfy medicinal chemists by doing this on the fly continues to be difficult.

The rapidly expanding field of deep generative molecular design has spurred new developments in drug discovery, including the synthesis and optimization of macromolecules and small compounds. However, there are still obstacles in the way of implementing and utilizing AI technologies, such as model limitations or lack thereof, accessibility issues, and a lack of high-quality data. To fully realize the promise of deep learning, intelligent generative models must be further developed and tested in authentic real-world drug development scenarios. With these advancements, the intelligent generative model paradigms could go from theoretical study to the development of medicines in practice and offer chemists and chemistry modelers simple-to-use toolkits for their everyday work.

The most interesting advance in medication delivery, repurposing, and discovery technologies is ML combined with bioinformatics tools. Due to its cost- and time-saving capabilities, ML is possible to improve biological data processing, drug discovery, and preclinical investigations. By combining machine learning with bioinformatics, standard, dependable, and repeatable computational approaches have been developed to meet the goals.

References

Agatonovic-Kustrin, S., & Morton, D. (2016). Data mining in drug discovery and design. In *Artificial neural network for drug design, delivery and disposition* (pp. 181-193). Academic Press.

Al Qaraghuli, M. M., Alzahrani, A. R., Niwasabutra, K., Obeid, M. A., & Ferro, V. A. (2017). Where traditional drug discovery meets modern technology in the quest for new drugs. *Annals of pharmacology and pharmaceutics*, *2*(11), 1-5.

Alvarez-Uria, G. (2014). Description of the cascade of care and factors associated with attrition before and after initiating antiretroviral therapy of HIV-infected children in a cohort study in India. *PeerJ*, *2*, e304.

Angermueller, C., Pärnamaa, T., Parts, L., & Stegle, O. (2016). Deep learning for computational biology. *Molecular systems biology*, *12*(7), 878.

Arden, N. S., Fisher, A. C., Tyner, K., Lawrence, X. Y., Lee, S. L., & Kopcha, M. (2021). Industry 4.0 for pharmaceutical manufacturing: Preparing for the smart factories of the future. *International Journal of Pharmaceutics*, *602*, 120554.

Badkul, A., Xie, L., Zhang, S., & Xie, L. (2024). TrustAffinity: accurate, reliable, and scalable out-of-distribution protein-ligand binding affinity prediction using trustworthy deep learning. *bioRxiv*, 2024-01.

Begam, B. F., & Kumar, J. S. (2012). A study on cheminformatics and its applications on modern drug discovery. *Procedia Engineering*, *38*, 1264-1275.

Berdigaliyev, N., & Aljofan, M. (2020). An overview of drug discovery and development. *Future medicinal chemistry*, *12*(10), 939-947.

Camacho, D. M., Collins, K. M., Powers, R. K., Costello, J. C., & Collins, J. J. (2018). Next-generation machine learning for biological networks. *Cell*, *173*(7), 1581-1592.

Chakravarty, K., Antontsev, V., Bundey, Y., & Varshney, J. (2021). Driving success in personalized medicine through AI-enabled computational modeling. *Drug Discovery Today*, *26*(6), 1459-1465.

Chen, H., Engkvist, O., Wang, Y., Olivecrona, M., & Blaschke, T. (2018). The rise of deep learning in drug discovery. *Drug discovery today*, *23*(6), 1241-1250.

Collins, F. S., & Varmus, H. (2015). A new initiative on precision medicine. *New England journal of medicine*, *372*(9), 793-795.

Deore, A. B., Dhumane, J. R., Wagh, R., & Sonawane, R. (2019). The stages of drug discovery and development process. *Asian Journal of Pharmaceutical Research and Development*, *7*(6), 62-67.

Dudley, J. T., Deshpande, T., & Butte, A. J. (2011). Exploiting drug-disease relationships for computational drug repositioning. *Briefings in Bioinformatics*, *12*(4), 303-311.

Goodfellow, I., Bengio, Y., Courville, A., & Bengio, Y. (2016). *Deep learning*, volume 1.

Herrling, P. L. (2005). The drug discovery process. *Imaging in drug discovery and early clinical trials*, 1-14.

Hinnerichs, T., & Hoehndorf, R. (2021). DTI-Voodoo: machine learning over interaction networks and ontology-based background knowledge predicts drug-target interactions. *Bioinformatics*, *37*(24), 4835-4843.

Hopkins, A. L. (2009). Predicting promiscuity. *Nature*, *462*(7270), 167-168.

Hughes, J. P., Rees, S., Kalindjian, S. B., & Philpott, K. L. (2011). Principles of early drug discovery. *British journal of pharmacology*, *162*(6), 1239-1249.

Jacob, L., & Vert, J. P. (2007). Kernel methods for in silico chemogenomics. *arXiv preprint arXiv:0709.3931*.

Jenkins, J. L., Bender, A., & Davies, J. W. (2006). In silico target fishing: Predicting biological targets from chemical structure. *Drug Discovery Today: Technologies*, *3*(4), 413-421.

Kempt, H., & Nagel, S. K. (2022). Responsibility, second opinions, and peer-disagreement: ethical and epistemological challenges of using AI in clinical diagnostic contexts. *Journal of Medical Ethics*, *48*(4), 222-229.

Kent, D. M., Steyerberg, E., & Van Klaveren, D. (2018). Personalized evidence based medicine: predictive approaches to heterogeneous treatment effects. *Bmj*, *363*.

Khan, S. R., Al Rijjal, D., Piro, A., & Wheeler, M. B. (2021). Integration of AI and traditional medicine in drug discovery. *Drug Discovery Today*, *26*(4), 982-992.

Lavecchia, A. (2015). Machine-learning approaches in drug discovery: methods and applications. *Drug Discovery Today*, *20*(3), 318-331.

Lavecchia, A., & Di Giovanni, C. (2013). Virtual screening strategies in drug discovery: a critical review. *Current medicinal chemistry*, *20*(23), 2839-2860.

LeCun, Y., Bengio, Y., & Hinton, G. (2015). Deep learning. *nature*, *521*(7553), 436-444.

Mak, K. K., & Pichika, M. R. (2019). Artificial intelligence in drug development: present status and future prospects. *Drug Discovery Today*, *24*(3), 773-780.

Mamoshina, P., Vieira, A., Putin, E., & Zhavoronkov, A. (2016). Applications of deep learning in biomedicine. *Molecular pharmaceutics*, *13*(5), 1445-1454.

Mishra, A. S., Vasanthan, M., & Malliappan, S. P. (2024). Drug Repurposing: A Leading Strategy for New Threats and Targets. *ACS Pharmacology & Translational Science*, *7*(4), 915-932.

Naylor, D. M., Kauppi, D. M., & Schonfeld, J. M. (2015). Therapeutic drug repurposing, repositioning, and rescue. *Drug Discovery*, *57*, 1-16.

Niazi, S. K. (2023). The coming of age of ai/ml in drug discovery, development, clinical testing, and manufacturing: The FDA perspectives. *Drug Design, Development, and Therapy*, 2691-2725.

Nicolaou, K. C. (2013). The emergence of the structure of the molecule and the art of its synthesis. *Angew. Chem. Int. Ed*, *52*(1), 131-146.

Nicolaou, K. C., & Montagnon, T. (2008). Molecules that changed the world. *Clinical Pharmacology and Pharmacokinetics*, *14*.

Pankevich, D. E., Altevogt, B. M., Dunlop, J., Gage, F. H., & Hyman, S. E. (2014). Improving and accelerating drug development for nervous system disorders. *Neuron*, *84*(3), 546-553.

Parvathaneni, V., Kulkarni, N. S., Muth, A., & Gupta, V. (2019). Drug repurposing: a promising tool to accelerate the drug discovery process. *Drug Discovery Today*, *24*(10), 2076-2085.

Preuer, K., Lewis, R. P., Hochreiter, S., Bender, A., Bulusu, K. C., & Klambauer, G. (2018). Deep Synergy: predicting anti-cancer drug synergy with Deep Learning. *Bioinformatics*, *34*(9), 1538-1546.

Roda, A., Guardigli, M., Pasini, P., & Mirasoli, M. (2003). Bioluminescence and chemiluminescence in drug screening. *Analytical and bioanalytical chemistry*, *377*, 826-833.

Sarkar, C., Das, B., Rawat, V. S., Wahlang, J. B., Nongpiur, A., Tiewsoh, I., Lyngdoh, N. M., Das, D., Bidarolli, M. & Sony, H. T. (2023). Artificial intelligence and machine learning technology-driven modern drug discovery and development. *International Journal of Molecular Sciences*, *24*(3), 2026.

Schmidhuber, J. (2015). Deep learning in neural networks: An overview. *Neural networks*, *61*, 85-117.

Schneider, G. (2018). Automating drug discovery. *Nature Reviews Drug Discovery*, *17*(2), 97-113.

Sherman, R. E., Anderson, S. A., Dal Pan, G. J., Gray, G. W., Gross, T., Hunter, N. L., LaVange, L., Marinac-Dabic, D., Marks, P. W., Robb, M. A. & Shuren, J. (2016). Real-world evidence—what is it and what can it tell us? *N Engl J Med*, *375*(23), 2293-2297.

Sinha, S., & Vohora, D. (2018). Drug discovery and development: An overview. *Pharmaceutical medicine and translational clinical research*, 19-32.

Topol, E. J. (2019). High-performance medicine: the convergence of human and artificial intelligence. *Nature Medicine*, *25*(1), 44-56.

Vamathevan, J., Clark, D., Czodrowski, P., Dunham, I., Ferran, E., Lee, G., Li, B., Madabhushi, A., Shah, P., Spitzer, M. & Zhao, S. (2019). Applications of machine learning in drug discovery and development. *Nature reviews Drug discovery*, *18*(6), 463-477.

Wallach, I., & Heifets, A. (2018). Most ligand-based classification benchmarks reward memorization rather than generalization. *Journal of chemical information and modeling*, *58*(5), 916-932.

Yun, Y. H., Lee, B. K., & Park, K. (2015). Controlled Drug Delivery: Historical perspective for the next generation. *Journal of Controlled Release*, *219*, 2-7.

Zhang, L., Tan, J., Han, D., & Zhu, H. (2017). From machine learning to deep learning: progress in machine intelligence for rational drug discovery. *Drug discovery today*, *22*(11), 1680-1685.

Zitnik, M., Nguyen, F., Wang, B., Leskovec, J., Goldenberg, A., & Hoffman, M. M. (2019). Machine learning for integrating data in biology and medicine: Principles, practice, and opportunities. *Information Fusion*, *50*, 71-91.

Chapter 14

Revolutionizing Drug Discovery: Generative Models and Reinforcement Learning in *De Novo* Drug Design

Tarun Madan Kanade[1,*]
and Radhakrishna Batule[2]
[1]Symbiosis Institute of Operations Management, Nashik, India
[2]Faculty of Management, Vishwakarma University, Pune, India

Abstract

The pharmaceutical industry is transforming because of the incorporation of artificial intelligence (AI) into the processes of drug development. Generative models and reinforcement learning (RL) play crucial roles in the process of de novo drug creation. The purpose of this chapter is to investigate the synergy that exists between these sophisticated AI approaches and their use in the process of identification of innovative medication candidates. Instruments include generative models, which include Generative Adversarial Networks (GANs) and Variational Autoencoders (VAEs), among others with similar names.

In the beginning of the chapter, an overview of the process of drug development is presented, with an emphasis placed on the limitations of conventional techniques and the need for innovative approaches. Following that, it delves into the principles of generative models and reinforcement learning, providing readers with a good core understanding of the subject matter. This article provides a comprehensive description of the integration of various technologies,

* Corresponding Author's Email: tarunbm45@gmail.com

In: Harnessing AI for Drug Discovery
Editor: Kolla Bhanu Prakash
ISBN: 979-8-89530-540-9

along with examples of frameworks that make use of the combined qualities of these technologies.
Several examples illustrate how AI-driven methodologies have led to significant advancements in the process of finding novel medicine candidates. These examples also exhibit successful applications. In addition, the chapter analyzes the challenges that this field is confronted with, such as the quality of the data, the requirements for computers, and the ethical implications, and it offers some suggestions for future study and development. It describes how the use of generative models and RL in drug design might have practical ramifications, including the tools, platforms, and data requirements that are required. The conclusion emphasizes the transformative potential of these technologies in the field of drug discovery, which opens the door to the creation of treatment programs that are both more efficient and successful.

Keywords: artificial intelligence, drug discovery, generative models, reinforcement learning, de novo drug design, GANs, VAEs, molecular optimization, pharmaceutical innovation, and AI in healthcare

Introduction

Many different businesses are being revolutionized by artificial intelligence (AI), and the pharmaceutical industry is no exception. Drug development, which has traditionally been a process that is both time-consuming and expensive, has seen a significant boost in both its efficiency and effectiveness because of the application of AI techniques. The use of machine learning (ML) algorithms makes it feasible to investigate massive datasets to identify potential medication candidates, make predictions about their characteristics, and maximize the effectiveness of these possibilities. These breakthroughs reduce the amount of time and money required to bring innovative medications to market, therefore solving significant issues that have been plaguing the business.

Importance of Generative Models and Reinforcement Learning (RL) in De Novo Drug Design

In the realm of artificial intelligence, approaches such as reinforcement learning (RL) and generative models have emerged as powerful strategies for the design of new drugs. Generative models, which include Generative Adversarial Networks (GANs) and Variational Autoencoders (VAEs), can produce new molecular structures by learning the underlying patterns of chemical compounds that are already in existence. The purpose of these models is to examine huge chemical areas, intending to identify new molecules that possess desirable characteristics that are typically beyond the grasp of conventional methods. The addition of reinforcement learning to this process brings about the optimization of these molecules via a method of trial and error. This methodology makes use of input from a simulated environment to repeatedly improve the chemical structures. Incorporating generative models with reinforcement learning (RL) speeds up the process of developing medicine candidates that are both effective and safe, pushing the boundaries of innovation in the pharmaceutical industry. (Martinelli, 2022)

Objectives and Scope of the Chapter

This chapter aims to provide a comprehensive overview of how generative models and reinforcement learning are revolutionizing de novo drug design. It will:

- The first step is to provide the principles of generative models and RL, elaborating on their functions and procedures in the process of drug development.
- Provide specifics on the incorporation of various technologies, focusing on frameworks that capitalize on the synergistic effects of their combined power.
- To demonstrate the practical effect and accomplishments of AI-driven medication design, it is necessary to provide case studies and applications that are based in the real world.
- Discuss the difficulties and constraints, such as the quality of the data, the complexity of the computations, and the ethical concerns involved.
- In this rapidly developing subject, it is important to investigate potential future paths and prospects for research and development.

Through an examination of these subjects, the chapter intends to highlight the revolutionary potential of artificial intelligence in the field of drug discovery. It will highlight the ways in which generative models and RL are paving the way for the creation of novel therapies that are both more efficient and effective.

Background and Foundations

The process of discovering and developing new drugs is a complex and multi-step procedure that often takes more than a decade to complete and requires a significant financial investment. Initially, the process begins with target identification, which is when researchers find a biological target that is associated with a disease. After that, the next stage is the hit identification process, which entails sorting through enormous chemical libraries in search of potential treatment candidates, also known as "hits." To increase their efficacy, safety, and pharmacokinetics, these hits make their way through the process of lead optimization, which involves refining their chemical structures. After that, the leads that have been refined are put through preclinical testing, which consists of both in vitro (using a test tube or cell culture) and in vivo (using an animal) studies to assess the biological activity and toxicity of the leads. Successful candidates are allowed to participate in clinical trials, which are conducted in three phases to evaluate the drug's effectiveness and safety in human subjects. In conclusion, after the drug has been approved by the regulatory body, it is then allowed into the market, and post-marketing monitoring is conducted to assess its effects over the long term (Greig, 2017).

Introduction to Generative Models: GANs, VAEs, and Other Relevant Architectures

One sort of machine learning model is known as a generative model, and it is responsible for producing new data samples by using distributions that have been learned in the past. With the use of these models, medication designers can develop novel molecular structures that have the desired characteristics. Generative models come in a variety of forms, two of which are GANs and VAEs.

- *GANs:* With adversarial processes, two neural networks, a generator and a discriminator, are trained simultaneously. The generator is responsible for the generation of the new data samples, which are then analyzed by the discriminator to determine whether they are legitimate about the actual data. Over time, the generator becomes more capable of producing realistic samples that the discriminator is unable to separate from actual data. When it comes to the creation of drugs, GANs can generate one-of-a-kind chemical structures that are comparable to known compounds that possess desired biological features (Zinchenko, 2024)
- *VAEs* are probabilistic models that encode and decode input data in a latent space. ii. Vaes are used to predict the future. Using this encoding-decoding process, VAEs can generate new data samples by pulling from the latent space. Because they build diverse molecular structures by investigating different portions of the latent space, VAEs are very useful in the process of drug synthesis. These sections correspond to a wide range of chemical properties and activities (Shafkat, 2018).

Other related structures include recurrent neural networks (RNNs) and transformer models, which are often used for sequence-based generative tasks such as the synthesis of peptide sequences and the prediction of chemical reactions.

Fundamentals of Reinforcement Learning: Q-Learning, DQNs, and Policy Gradients

Reinforcement learning, sometimes known as RL, is a method of machine learning in which an agent learns to make choices by interacting with its surroundings. Depending on the tasks it does, the agent is rewarded or punished, to maximize the cumulative benefits over time.

- *Q-learning*: A model-free reinforcement learning technique in which the agent is taught a Q-value function to estimate the rewards that are expected to be received for certain behaviors in a given state. To make decisions that maximize the long-term advantages, the agent makes use of this Q-value function. It is possible to apply Q-learning in the process of drug design to iteratively analyze and enhance the characteristics of molecular structures (Meng Zhang, 2023).

- *Deep Q-networks (DQNs):* Through the use of deep neural networks, DQNs are able to approximate the Q-value function, which enables agents to manage complicated state spaces such as those seen in molecular design. Direct quantum networks (DQNs) enable the agent to explore and optimize complex chemical environments in a more efficient manner (Singh, 2023).
- *Policy Gradients*: Instead of optimizing the Q-value function, policy gradient techniques focus on optimizing the agent's strategy for action selection. Gradients are used in these methods in order to modify the policy parameters in a manner that results in an increase in the anticipated reward. Policy gradients are helpful in the process of drug design because they make it possible to optimize sequences or molecular structures that have a large and continuous action space.

By gaining an understanding of these basic concepts, researchers can use generative models and reinforcement learning to foster innovation and speed up the process of drug discovery, which ultimately results in the production of innovative medicines that are more efficient and effective.

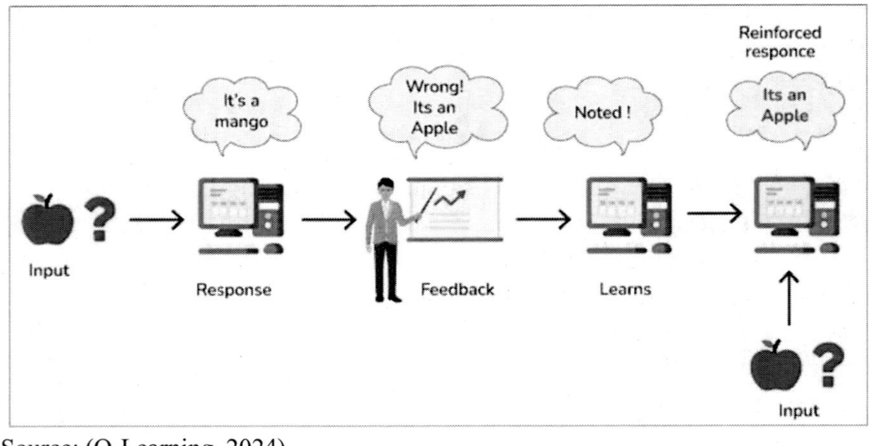

Source: (Q-Learning, 2024).

Figure 1. Q-learning in Reinforcement Learning.

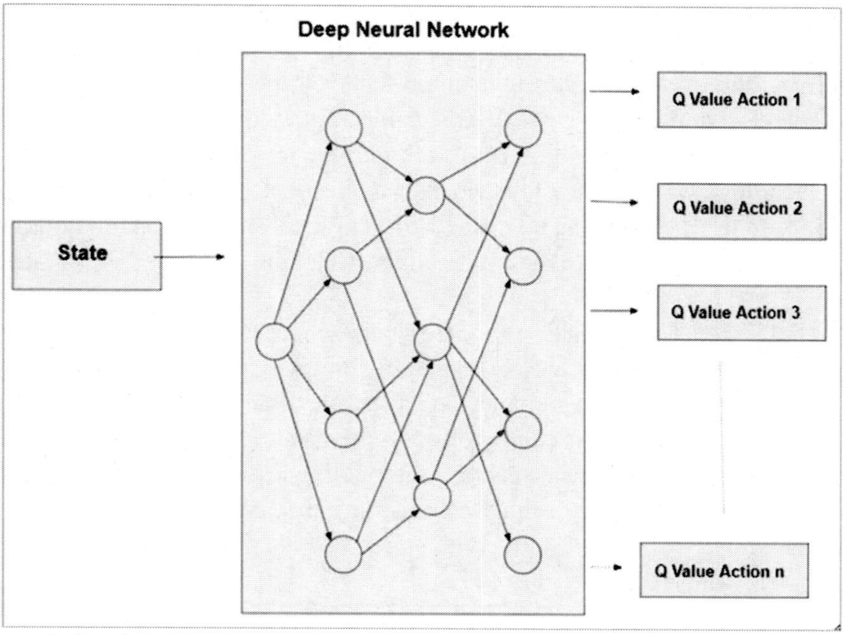

Source: (Deep Q-Learning, 2023).

Figure 2. Deep Q-Learning Network.

Generative Models in Drug Design

In the field of drug development, generative models play a significant role since they make it possible to generate novel chemical structures that have the potential to be effective medicines. Traditional approaches to drug discovery usually depend on screening pre-existing chemical libraries, which may be both time-consuming and restricting in terms of the breadth of the possible applications. The use of machine learning algorithms in generative models, on the other hand, allows for the exploration of huge chemical areas and the creation of new molecular entities that possess desired characteristics. This results in an increase in the diversity and originality of candidate molecules.

Generative adversarial networks, also known as GANs, and variational autoencoders, often known as VAEs, are two types of generative models that are widely used in this application. Two neural networks, a generator and a discriminator, are used in the construction of GANs. These neural networks are trained concurrently via the use of an adversarial process. First, the generator creates new chemical structures, and then the discriminator determines whether or not those structures are genuine. During its operation, the generator will gradually improve its capacity to generate new molecules that are both realistic and indistinguishable from those of recognized compounds. On the other hand, VAEs can learn how to encode molecules in a latent space before decoding them back into the molecular space. Utilizing this encoding-decoding strategy, VAEs can construct unique compounds by taking samples from the latent space and doing research on separate places that exhibit chemical properties and activities that are connected (Parakatta, 2023).

These models can make use of domain-specific information, such as chemical laws and data on biological activity, to direct the generation process in the direction of the development of molecules that are similar to drugs. Generative models are particularly effective in discovering novel compounds that possess certain qualities, such as high binding affinity, low toxicity, and favorable pharmacokinetic features. This is since they can tailor the generation process.

Case Studies: Successful Generative Model Applications in Drug Discovery

Several case studies demonstrate the successful use of generative models in drug development.

1. The development of small-molecule inhibitors for DDR1 kinase was accomplished by Insilico Medicine via the use of Generative Adversarial Networks (GANs). In a span of twenty-one days, the model was able to produce thirty thousand novel structures, with some compounds exhibiting considerable binding affinity during the subsequent experimental validation. In this particular case, the effectiveness and potential of GANs in accelerating the first stages of drug discovery are brought to light (Zheng, 2019).
2. *GENTRL, developed by Insilico Medicine*: GENTRL, which is a generative tensorial reinforcement learning model, was integrated into the process of developing new inhibitors for DDR1 kinase. Only 46 days were required for the model to develop interesting molecules, which is a considerable reduction from the average amount of time required for drug discovery. The fact that several of these drugs

demonstrated nanomolar activity in biological testing demonstrates the practical usefulness of generative models (Alex Zhavoronkov, 2019).

3. *Variational Autoencoders for Antibiotic Discovery:* To come up with new antibiotic candidates, a research team used VAEs. The model can develop novel compounds with structural features that are comparable to those of known antibiotics if it is trained on a dataset of antibiotics. In laboratory experiments, several of the compounds that were created displayed excellent antibacterial action. This highlights the promise of VAEs in the discovery of novel medicines in the face of developing antibiotic resistance (Marcelo C. R. Melo, 2021).

Challenges and Limitations of Using Generative Models

Despite their promise, generative models in drug design face several challenges and limitations:

1. The success of generative models is strongly dependent on the quality and amount of the training data. Therefore, it is important to consider both of these factors. The production of molecules that are either unrealistic or inefficient might be caused by data that is inaccurate, biased, or inadequate information. It is of the utmost importance to guarantee high-quality datasets, yet this might be difficult to do owing to the confidential nature of pharmaceutical data.
2. *The interpretability of models is a common concern.* Generative models, especially those that are built on deep learning, are sometimes referred to as "black boxes." Because it might be difficult to comprehend how these models arrive at judgments and produce certain compounds, the pharmaceutical sector, which is heavily regulated, has challenges in gaining confidence and implementing them. One of the current areas of study is aimed at improving the interpretability of models.
3. *Computational Resources*: Generative models need a substantial amount of computational resources, especially when dealing with large datasets and complex molecular structures. It is possible that many research institutions and small pharmaceutical enterprises

would have difficulties in gaining access to high-performance computer resources on their own.

4. *Validation and Experimental Testing*: Generative models can develop novel compounds in artificial intelligence; nevertheless, to determine their biological activity and drug-like characteristics, they need to be validated by experimental testing. In addition to being time-consuming and expensive, this validation approach is not going to be successful for all of the chemicals that are created in the experimental testing.
5. *Exploration of Chemical Space*: Although generative models are capable of covering an extensive region, the limitations and biases of these models may result in opportunities being lost. The challenge of ensuring that the chemical space is thoroughly investigated continues to be a significant one (Xiangxiang Zeng, 2022).

Generative models have the potential to change drug design by successfully developing novel chemical structures. In conclusion, this is the promise of generative models. However, to fully leverage their advantages in drug development, it is essential to solve obstacles such as the quality of the data, the interpretability of the model, the computing requirements, and the confirmation of the experimental results.

Reinforcement Learning: Concepts and Techniques

In the field of machine learning, Reinforcement Learning (RL) refers to a kind of learning in which an agent learns to make choices by interacting with its surroundings. In contrast to supervised learning, which makes use of data that has been labeled, reinforcement learning (RL) is based on a feedback loop in which the agent is rewarded or punished depending on the behavior it exhibits. The primary goal is to have an understanding of a policy that maximizes the accumulation of benefits over a while. The way that individuals learn is quite similar to this process, which is via trial and error.

RL's primary components are:

- Agent: The decision-maker who interacts with their surroundings.
- Environment: The system or setting in which the agent functions.
- State: Describes the current environment.

- Action: A move or choice made by the agent that impacts the state.
- Reward: Feedback from the environment on the agent's actions.
- Policy: The agent's method for determining actions based on current conditions.
- Value Function: Calculates the predicted cumulative reward from a given condition.

The agent's goal is to establish an optimal policy that maximizes the predicted cumulative reward by experimenting with various behaviors and learning from the results.

Types of Reinforcement Learning Algorithms Used in Drug Design

Several RL algorithms are particularly useful in drug design:

- *Q-Learning*: A model-free technique in which the agent learns a Q-value function to estimate the anticipated reward for an action. This is done in a specific state. The Q-values are adjusted by the agent depending on the rewards that have been obtained as well as the benefits that are anticipated to be earned in the future. When used in discrete action spaces, this method is both straightforward and efficient.
- *Deep Q-Networks*, also known as DQNs, are an extension of Q-learning that makes use of deep neural networks in order to calculate an approximation of the Q-value function. The agent is able to efficiently manage high-dimensional state and action spaces as a result of this. Due to their ability to optimize complex chemical structures, DQNs are very useful in the process of drug creation.
- *Policy Gradient approaches*: These methods optimize policies by altering parameters to maximize the benefits that are anticipated. In contrast to value-based techniques such as Q-learning, policy gradients are well-suited for continuous and expansive action spaces. As a result, they are suitable for molecular design issues that include a wide range of activities.

A value-based and policy-based approach are both included in the actor-critical approaches, which are the fourth kind. In response to the criticism provided by the critic, the actor makes adjustments to the policy. The critic evaluates the policy by formulating estimates of value functions. This combination makes it possible to learn in a more consistent and efficient manner, especially in instances when the context is complex.

Reward Functions and Their Significance in Molecular Optimization

When it comes to reinforcement learning (RL), reward functions are essential since they set objectives and guide the learning process of the agent. The reward function is a component of molecular optimization that is designed to convey the degree to which certain molecular characteristics are appealing. It is essential to the process of successful drug development to develop a reward function that offers effective benefits.

Important features of reward functions in molecular optimization include:

1. *Binding Affinity:* Molecules are often evaluated according to their anticipated binding affinity to a target protein. A high binding affinity indicates a stronger interaction with the target, which is often sought in drug design.
2. *Drug-likeness*: Rewards may include Lipinski's Rule of Five, which assesses compounds according to molecular weight, hydrogen bond donors, and acceptors.
3. *Toxicity and Safety*: By imposing negative incentives on substances with elevated toxicity or adverse effects, agents may circumvent potentially hazardous candidates.
4. *Synthetic Accessibility*: Molecules that are readily synthesizable are preferred. Rewards may be contingent upon parameters that evaluate the viability of chemical synthesis.
5. *Diversity:* The synthesis of various compounds facilitates greater exploration of chemical space. Rewards may include sanctions for developing chemicals that closely resemble those already evaluated.

Reinforcement learning provides a robust foundation for enhancing molecular architectures in pharmaceutical design. Researchers can effectively guide AI models in identifying novel and potent drug candidates by using diverse reinforcement learning algorithms and meticulously designing incentive mechanisms. The ability to incrementally adjust compounds via feedback-based optimization renders reinforcement learning a valuable asset in contemporary pharmaceutical research.

Integration of Generative Models and Rl

How Generative Models and RL Complement Each Other in Drug Design

Generative models and reinforcement learning (RL) are two powerful AI methodologies that, when integrated, provide a synergistic strategy for drug development. Generative models, like GANs and VAEs, are proficient at creating novel chemical structures by discerning patterns from existing datasets. They explore extensive chemical domains, producing a varied array of potential treatment possibilities. Nonetheless, these models may not inherently optimize for certain attributes like efficacy or safety.

This is where reinforcement learning is applicable. Reinforcement learning is designed to be refined via trial and error, using a process for the repeated enhancement of synthesized molecules. By establishing specific incentive functions, RL agents may guide generative models to produce molecules that are both novel and exhibit favorable attributes, like high binding affinity, low toxicity, and optimal pharmacokinetics. Consequently, generative models provide diverse initial frameworks, while reinforcement learning enhances these structures to meet stringent drug design criteria, culminating in a robust and efficient methodology (Nilesh, 2024).

Frameworks and Methodologies for Integrating These Technologies

Integrating generative models with RL requires multiple frameworks and procedures, each designed to optimize the strengths of both approaches:

1. *Sequential Generation and Optimization*: This method consists of two distinct phases. Initially, a generative model (such as a GAN or VAE) produces a substantial quantity of novel molecules. These molecules are then sent to a reinforcement learning agent, which evaluates and enhances them using established reward functions. This technique

maintains initial diversity while focusing on optimal candidates via reinforcement learning (Manan Goel, 2021).

2. *Reinforcement Learning-Enhanced Generation*: This approach incorporates the generative model directly into the reinforcement learning framework. The RL agent provides instantaneous input to the generative model, therefore dynamically modifying the generation process. This creates a closed-loop system where each synthesized molecule is promptly assessed and used to refine the generative model, facilitating more efficient convergence on optimal structures (Hannes H. Loeffler, 2024).
3. Hybrid models integrate generative and reinforcement learning techniques into a unified framework. A policy gradient method may be used to modify the parameters of a VAE, integrating generation and optimization into a singular process. Hybrid models may streamline the process, reducing computing demands and perhaps improving the quality of synthesized molecules.(Masatoshi Uehara, 2024).

Examples of Integrated Systems in Current Research

1. *GENTRL by Insilico Medicine:* GENTRL exemplifies the use of generative tensorial reinforcement learning in the creation of new molecules. This approach combines generative models with reinforcement learning to generate candidates for DDR1 kinase inhibitors. GENTRL showed the capability to produce promising compounds at a markedly accelerated pace compared to conventional approaches, underscoring the practical advantages of this integrated approach (Zhavoronkov, 2019).
2. *MoleculeRL:* IBM created MoleculeRL, which enhances molecules for designated attributes via Variational Autoencoders and Reinforcement Learning. The VAE generates initial molecules, which are then refined by a reinforcement learning agent. This method has proven useful in generating drug-like molecules with high affinity for target proteins (Hwanhee Kim, 2022).
3. *ReLeaSE (Reinforcement Learning for Structural Evolution):* ReLeaSE integrates a deep generative model with a reinforcement learning framework to produce innovative chemical structures. The generative model suggests novel compounds, while the reinforcement

learning component assesses them according to a reward function that encompasses drug-likeness, synthetic feasibility, and biological activity (Sara Romeo Atance, 2022).

The integration of generative models with reinforcement learning is a pioneering methodology in drug development, merging the inventiveness of generative models with the precision of reinforcement learning. This collaboration facilitates the rapid discovery and enhancement of new drug candidates, significantly accelerating the drug development process and improving the probability of uncovering effective new therapies.

Case Studies and Applications

Real-World Examples of De Novo Drug Design Using Generative Models and RL

Insilico Medicine's GENTRL (Generative Tensorial Reinforcement Learning) system exemplifies the integration of generative models with reinforcement learning (RL) in medication development. GENTRL employs a generative model to create distinctive chemical compounds and then use reinforcement learning to optimize these molecules for defined objectives. Insilico used GENTRL to create inhibitors for the DDR1 kinase, an enzyme associated with cancer proliferation. The method yielded promising molecules in 21 days, significantly reducing time and costs compared to prior techniques (Amit Gangwal, 2024).

Atomwise

Atomwise, a leader in AI-driven drug discovery, developed AtomNet, a deep-learning network trained on millions of chemical compounds. AtomNet, while not exclusively a reinforcement learning model, utilizes deep convolutional neural networks to forecast bioactivity and enhance molecules, illustrating the practical use of AI in drug development. Atomwise collaborated with researchers to develop novel inhibitors for several diseases, including Ebola and multiple sclerosis, showcasing the scalability and efficacy of AI in practical drug discovery (Izhar Wallach, 2015).

BenevolentAI employs generative models and reinforcement learning to accelerate medication development. Their approach use machine learning algorithms to sift through extensive biological data and identify potential treatment possibilities. BenevolentAI identified Baricitinib, a drug originally developed for rheumatoid arthritis, as a potential treatment for COVID-19. This discovery, leading to clinical trials, illustrates AI's capability to swiftly repurpose current drugs in reaction to emerging health crises (Hodos, 2021).

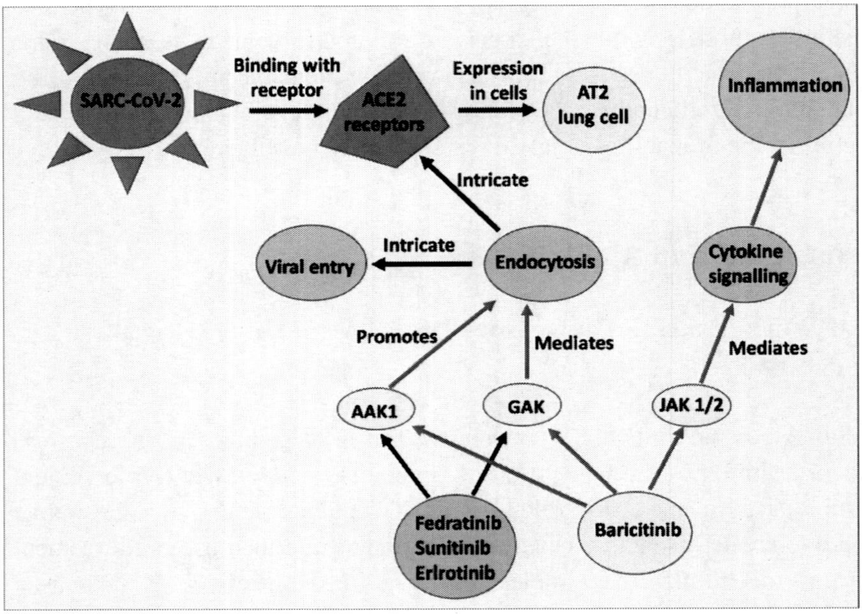

Source: (Atul Sharma, 2020).

Figure 3. BenevolentAI knowledge graph integrating biomedical data from existing structured and unstructured sources.

Success Stories and Breakthroughs

Insilico Medicine's research on DDR1 kinase inhibitors in collaboration with GENTRL has achieved significant progress. The AI system generated novel molecules, many of which had significant binding affinity in laboratory evaluations. This accomplishment not only validated the viability of integrating generative models with reinforcement learning but also showcased the rapidity and efficacy of AI-facilitated medication development (Alex Zhavoronkov Y. A., 2019).

Exscientia

DSP-1181

Exscientia partnered with Sumitomo Dainippon Pharma to create DSP-1181, a potential medication for obsessive-compulsive disorder. Exscientia significantly decreased the duration of drug discovery by the use of AI-driven generative models. DSP-1181 progressed from initial concept to preclinical evaluation in 12 months, much less than the typical timeframe. This outcome illustrates AI's capacity to accelerate the drug development process (Burki, 2020).

Lessons Learned and Best Practices

1. *Data Quality and Integration*: The training of generative models and reinforcement learning agents needs high-quality, meticulously managed datasets. Integrating several data sources, including chemical, biological, and clinical data, enhances the model's ability to generate and refine potential drug candidates.
2. *Interdisciplinary Collaboration*: Successful AI-driven medication development requires strong cooperation among AI specialists, chemists, and biologists. This interdisciplinary approach ensures that AI models are informed by domain-specific information, hence enhancing their relevance and usefulness.
3. *Iterative Testing and Validation*: Ongoing iteration between in-silico predictions and experimental validation is crucial. This feedback mechanism enhances the accuracy and reliability of AI models progressively. Real-world validation ensures that the synthesized compounds meet the requisite biological and chemical requirements.
4. *Ethical and Regulatory Considerations*: Comprehending the ethical and regulatory framework is essential. Organizations must ensure that their AI algorithms are transparent and comply with regulatory standards, particularly in clinical trials and patient safety.
5. *Scalability and Computational Resources*: Utilizing cloud computing and high-performance computing resources may address the computational requirements for training generative models and reinforcement learning algorithms. Scalable infrastructure facilitates more extensive and intricate simulations, hence enhancing the discovery process.

These case studies and best practices illustrate the significance and potential of integrating generative models with reinforcement learning in drug development. The successes of companies like Insilico Medicine, Atomwise, and Exscientia illustrate the transformative potential of AI-driven methodologies, while the insights acquired provide a blueprint for future progress in this rapidly evolving sector.

Challenges and Future Directions

Current Challenges in Using Generative Models and RL in Drug Design

- Data Quality and Availability: A primary challenge is the accessibility and integrity of data. Generative models and reinforcement learning algorithms need extensive, diverse, and high-quality datasets for optimal performance. Nonetheless, data inside the pharmaceutical business is often proprietary, disjointed, and may include inaccuracies and biases. The absence of standardized data may lead to models that are less generalizable and effective.
- Training generative models and reinforcement learning agents necessitates substantial computational resources. These models often need substantial computational resources, including high-performance computing (HPC) environments and cloud-based infrastructures. Small enterprises and academic institutions may have challenges in acquiring these resources, consequently limiting their capacity for innovation.
- Model interpretability is a substantial challenge for AI models. Numerous generative models and reinforcement learning algorithms, especially those using deep learning, function as "black boxes," rendering it infeasible to understand their decision-making processes. The absence of transparency may erode trust and acceptance in the highly regulated pharmaceutical industry, where comprehending the rationale behind drug design choices is essential.
- Integrating AI-driven methodologies with conventional drug development processes may be arduous. Pharmaceutical companies must modify their processes to effectively integrate AI technology,

requiring enhancements in infrastructure, scientific training, and novel data management and analysis methodologies.

Ethical Considerations and Potential Risks

- *Bias and Fairness*: AI models can perpetuate biases seen in training data. In medication design, this might imply that specific illnesses or patient populations are underserved. To avoid aggravating health inequities, AI systems must be fair and unbiased.
- *Transparency and Accountability*: The use of AI in medication development engenders apprehensions pertaining to openness and accountability. Decisions made by AI systems must be explicable, and there should be explicit criteria delineating accountability for AI-generated outcomes, especially in healthcare settings.
- *Patient Privacy:* AI models often need access to confidential health information. Safeguarding patient privacy and adhering to regulations such as GDPR and HIPAA is essential for preserving trust and preventing legal repercussions.

Future Trends and Opportunities for Research and Development

- *Improved Data Sharing and Collaboration*: Collaboration across institutions and countries may significantly improve datasets for training generative models and reinforcement learning algorithms. Public-private partnerships and data consortiums may facilitate the aggregation of resources and data, leading to enhanced model performance and innovation.
- *Advanced Interpretability Techniques*: Investigations toward enhancing the interpretability of AI models are ongoing. Explainable AI (XAI) methodologies enhance model transparency, enabling researchers to comprehend and have confidence in AI-generated therapeutic opportunities. Developing methodologies to analyze and understand the decision-making processes of generative models and reinforcement learning agents will be essential.
- Future advancements include the integration of multi-omics data (genomics, proteomics, and metabolomics) to get a comprehensive

understanding of disease etiology and therapeutic responses. This holistic approach may provide more efficient and customized drug designs.

- *Continuous Learning Systems:* The advancement of AI systems capable of learning and adapting to new data is a significant milestone. These systems may remain abreast of the latest research findings and clinical data, enhancing their predictive capability and utility over time.
- Establishing robust ethical frameworks and regulatory standards for AI in drug research is essential. These frameworks must tackle issues of bias, justice, accountability, and transparency, guaranteeing that AI-generated medications benefit all patient demographics equitably.

Despite significant challenges, the integration of generative models and reinforcement learning in drug development has substantial promise. The pharmaceutical industry may use AI to revolutionize drug research and development by addressing current difficulties and seizing future opportunities.

Practical Considerations and Implementation

Tools and Platforms for Implementing Generative Models and RL in Drug Design

Software and Frameworks

Several software tools and frameworks are available for incorporating generative models and reinforcement learning (RL) into drug development. Key platforms include:

1. DeepChem is an open-source software suite that utilizes deep learning in the fields of chemistry and biology. It encompasses instruments for molecular synthesis, property forecasting, and data administration.
2. TensorFlow and PyTorch are prominent deep learning frameworks that enable the development and training of generative models, including GANs, VAEs, and reinforcement learning algorithms. They

provide extensive libraries and pre-constructed modules to facilitate model development.

3. RDKIt is an invaluable toolkit for molecular modeling and cheminformatics applications. It seamlessly integrates with deep learning frameworks for the preprocessing and analysis of chemical data (Venkata Chandrasekhar, 2023)
4. Chainer Chemistry: This enhancement to the Chainer deep learning framework facilitates molecular representation and manipulation, hence streamlining the integration of generative models with chemical data (Seiya Tokui, 2019)
5. Reinforcement Learning Libraries: Libraries such as OpenAI Gym and Stable Baselines provide implementations of several reinforcement learning algorithms and environments, facilitating the construction and evaluation of reinforcement learning models for drug discovery.

Cloud Platforms and High-Performance Computing

Cloud platforms like Amazon Web Services (AWS), Google Cloud Platform (GCP), and Microsoft Azure provide scalable resources for the training and deployment of artificial intelligence models. They provide specialized machine learning services, like AWS SageMaker, GCP AI Platform, and Azure Machine Learning, which facilitate the extensive calculations necessary for generative models and reinforcement learning algorithms.

Data Requirements and Preprocessing Techniques

1. *Data Quality and Quantity*: High-quality, diverse datasets are essential for training effective generative models and reinforcement learning agents. Examples of drug design factors include chemical structures, biological activity data, pharmacokinetic and pharmacodynamic characteristics, and clinical trial results. Ensuring data quality requires comprehensive validation, curation, and preparation.
2. *Preprocessing Techniques:*
    - *Normalization and Standardization*: Chemical data needs normalization to guarantee that features are on a uniform scale.

Standardization may include altering molecular properties to conform to certain distributions.

- *Molecular Representations*: Accurate molecular representation is crucial for AI models. Prevalent formats including SMILES (Simplified Molecular Input Line Entry System), InChI (International Chemical Identifier), and graph-based representations. Each format necessitates certain preparation protocols, such tokenization for SMILES strings or the construction of graphs for chemical representations.
- *Feature Engineering*: Deriving significant attributes from molecular data may enhance model efficacy. This may include the computation of molecular descriptors (e.g., molecular weight, logP) and fingerprints (e.g., ECFP, MACCS keys).
- *Data Augmentation*: To address data scarcity, techniques like data augmentation might provide more training instances. This may include altering existing molecules or generating new ones using generative models. (Dalwinder Singh, 2020)

Computational Resources and Scalability Issues

Training generative models and RL agents requires a significant number of resources. Key challenges include:

1. *High Computational Cost:* Training deep generative models and reinforcement learning agents necessitates substantial computing resources, particularly for extensive datasets and intricate models. This often requires access to GPUs or TPUs.
2. *Memory Requirements:* Handling extensive molecular datasets and model parameters requires considerable memory resources. Effective memory management and optimization techniques are essential for preventing bottlenecks.
3. *Parallel Processing:* Employing parallel processing may significantly reduce training duration. Distributed training frameworks, such TensorFlow's Distributed Strategy and PyTorch's Distributed Data Parallel, enable model training over several GPUs or nodes.

4. *Cloud and HPC Solutions:* Cloud services and high-performance computing (HPC) clusters may address scalability issues. Cloud systems provide adaptable, on-demand resources, while HPC clusters deliver the computational power necessary for intricate tasks. (LeCun, 2019)

Best Practices for Scalability

1. *Efficient Algorithm Design*: Developing efficient algorithms that minimize computational overhead and enhance resource efficiency is essential. This entails selecting appropriate model designs and optimization techniques.
2. *Incremental Learning:* Employing incremental learning methodologies enables models to assimilate new data without necessitating complete retraining, hence enhancing efficiency and scalability.
3. *Optimization Techniques*: Employing sophisticated optimization approaches, like gradient accumulation and mixed-precision training, may improve computing efficiency and model performance.
4. *Resource Management*: Efficient resource management entails overseeing and enhancing the use of computing resources, guaranteeing that hardware is employed effectively and economically. (Khader Hamdia, 2021)

The use of generative models and reinforcement learning in drug design involves using contemporary tools and platforms, ensuring proper data preparation, and tackling computational scalability. Researchers may optimize the potential of AI-driven medication development by effectively integrating these attributes.

Suggested Framework to the Industry

Implementing generative models and reinforcement learning (RL) in the pharmaceutical industry requires a comprehensive framework that addresses various critical parameters. Here's a suggested framework with nine key parameters:

Data Management and Integration

Ensuring high-quality data is essential for effective AI-driven medicine creation, necessitating stringent validation and curation processes. Standardized molecular data formats, such SMILES and InChI, must be used to guarantee consistency and reliability. Incorporating diverse data types, including chemical, biological, and clinical information, enhances datasets and elevates model accuracy and robustness. Utilizing a vast array of data sources, including public databases like ChEMBL and PubChem, as well as private datasets and collaborative efforts with academic institutions, enhances and enriches the data foundation.

Molecular Representation and Feature Engineering

Employing suitable molecular representation techniques, whether SMILES strings, molecular graphs, or fingerprints, is essential for enhancing AI models in drug development. To ensure precision and pertinence, each representation must be selected based on the specific application and model criteria. Moreover, using feature engineering to derive essential chemical descriptors and attributes enhances model input and predictive efficacy, making models more robust and effective in generating innovative drug candidates.

Model Selection and Development

Employing advanced generative models, like GANs, VAEs, and recurrent neural networks (RNNs), especially tailored for the generation of chemical data is essential for advancing drug research. These sophisticated models may generate diverse and distinctive chemical structures, expanding the spectrum of potential treatment options. Moreover, the selection of appropriate reinforcement learning (RL) algorithms, like Q-learning, DQNs, or policy gradients, is essential for optimizing molecular attributes and directing the generation process. These reinforcement learning algorithms enhance the synthesized molecules to meet certain criteria, guaranteeing that new pharmaceuticals are both effective and safe.

Computational Infrastructure

Addressing the computational requirements for training and deploying AI models in drug design necessitates the use of cloud platforms such as AWS, GCP, and Azure, in addition to high-performance computing (HPC) resources. These platforms provide scalable infrastructure capable of supporting the growing intricacy of data and models. Moreover, using parallel processing methods and distributed training frameworks is essential for accelerating model training and enhancing overall efficiency, leading to expedited and more effective drug discovery processes.

Model Training and Optimization

Enhancing model performance and generalizability necessitates the establishment of robust training procedures that include methodologies such as data augmentation, cross-validation, and hyperparameter tuning. Furthermore, using sophisticated optimization techniques like as gradient accumulation, mixed-precision training, and efficient batch processing enhances resource usage and reduces training duration, leading to more effective model training.

Evaluation and Validation

Establishing and tracking key performance measures, including binding affinity, drug-likeness, and synthetic accessibility, is essential for evaluating model outcomes. Ongoing evaluation of these indicators facilitates progress monitoring and identification of improvement prospects. Additionally, collaborating with experimental labs to validate AI-generated pharmaceutical candidates via in vitro and in vivo testing ensures their real-world relevance and reliability, therefore bridging the divide between computational forecasts and empirical outcomes.

Interdisciplinary Collaboration

Facilitating cooperation among AI specialists, chemists, biologists, and pharmacologists is essential for effective team integration. This interdisciplinary approach ensures that domain expertise and practical concerns feed models, leading to more precise and relevant outcomes. Moreover, promoting effective communication and providing training programs are essential for ensuring that all team members comprehend the AI tools and methodologies used, leading to a unified and proficient workforce.

Ethical and Regulatory Compliance

The effective design and use of AI models need the establishment and application of ethical standards that tackle issues of bias, transparency, and accountability. This ensures that patient safety and societal impacts are assessed at every level. Moreover, adherence to regulatory standards, including those set by the FDA and EMA, is essential for pharmaceutical development. Maintaining comprehensive documentation and validation records helps in regulatory submissions and clearances, ensuring that AI-driven drug development processes comply with all legal and ethical norms.

Continuous Improvement and Innovation

Iterative enhancements need ongoing feedback loops among model predictions, experimental data, and clinical results. This continuous input enhances predictions and elevates model precision. Moreover, investing in ongoing research and development is essential for investigating novel AI approaches, incorporating emerging technologies, and sustaining a competitive edge in drug discovery. These initiatives ensure that the medication development process remains innovative and progressive.

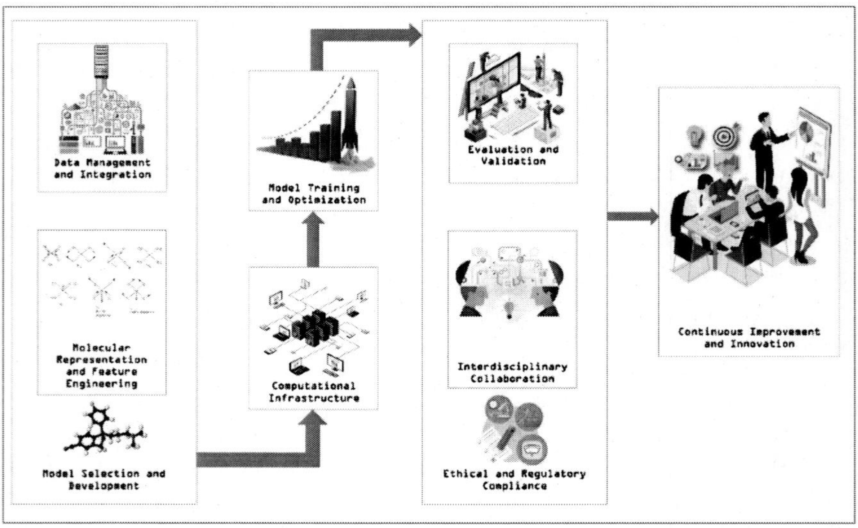

Designed by Dr Tarun Madan Kanade.

Figure 4. AI-Driven Drug Discovery Implementation Framework (ADDFI).

ADDFI enhances the drug development process by integrating methodologies from artificial intelligence, machine learning, and pharmaceutical sciences. The emphasis is on structured initial assessment, iterative pilot initiatives, scalable integration, continuous monitoring, and stakeholder engagement to facilitate the effective and successful use of generative models and reinforcement learning in drug design. This idea promotes flexible and cooperative approaches to pharmaceutical research and development, using technological advancements and interdisciplinary collaboration to foster innovation.

Implementation Roadmap

Early assessment and planning are essential for evaluating current abilities, identifying deficiencies, and establishing precise objectives for the use of generative models and reinforcement learning in drug design. Initiating pilot projects enables the validation of the selected methodologies' feasibility and effectiveness, while also refining the framework and demonstrating proof-of-concept. Effective pilot projects are expanded and incorporated into the broader drug development framework, guaranteeing that infrastructure, protocols, and personnel are prepared for extensive implementation. Performance must be consistently monitored and assessed, with key performance indicators (KPIs) defined to measure progress, quantify success, and implement necessary adjustments. Interacting with stakeholders, including regulatory bodies, industry collaborators, and patients, guarantees that the AI-driven drug development process aligns with overarching objectives and anticipations, fostering trust and cooperation.

By following this comprehensive approach, the pharmaceutical industry may effectively use generative models and reinforcement learning to transform drug design, leading to accelerated, more efficient, and innovative drug development processes.

Conclusion

This chapter examines the amalgamation of generative models and reinforcement learning (RL) in drug discovery, highlighting critical methodologies, challenges, and potential advancements. We started by discussing the principles of AI in drug discovery, emphasizing the significance of generative models and reinforcement learning in accelerating the identification and optimization of novel therapeutic options.

Critical challenges underscored the importance of data quality and diversity, essential for training efficient AI models. We highlighted the significance of molecular representation methods, including SMILES strings and molecular graphs, with enhanced feature extraction strategies, to augment model accuracy and predictive efficacy. Employing sophisticated generative models like as GANs and VAEs, with reinforcement learning approaches like Q-learning and policy gradients, is essential for improving molecular properties and regulating the generation process.

The Transformative Potential of Generative Models and Reinforcement Learning in Drug Discovery

Generative models and reinforcement learning signify a paradigm shift in drug discovery, with transformative potential throughout all stages of the drug development process. These AI-driven tools expedite the exploration of vast chemical spaces by generating new chemical structures and enhancing molecular properties, perhaps uncovering therapeutic opportunities that conventional approaches may overlook. The ability to predict and modify molecular structures according to certain pharmacological profiles accelerates the discovery process and enhances the likelihood of developing safer and more effective medications.

Moreover, the collaboration between generative models and reinforcement learning enables iterative learning and adaptation based on experimental data. This iterative approach enhances model efficacy and enables the creation of personalized treatments for certain patient populations or disease variants (Xiangxiang Zeng F. W.-g., 2022).

Final Thoughts on the Future Landscape of AI-Driven Drug Design

The future of AI-driven pharmaceutical design is auspicious but intricate. Progress in AI algorithms, processing capacity, and data accessibility will advance the subject. Ethical issues, including bias reduction and openness in AI decision-making, will be essential for maintaining patient safety and regulatory adherence. Collaboration among AI specialists, biologists, chemists, pharmacologists, and regulatory agencies will be essential.

In summary, despite existing challenges, the transformative impact of generative models and reinforcement learning in drug discovery is evident. The pharmaceutical business may alter therapeutic research and enhance patient outcomes by strategically and collaboratively using emerging technology.

References

Alex Zhavoronkov, Y. A. (2019). Deep learning enables rapid identification of potent DDR1 kinase inhibitors. *Nature Biotechnology, 37*, pages1038–1040. https://doi org/ https://doi.org/10.1038/s41587-019-0224-x.

Alex Zhavoronkov, Y. A. (2019). Deep learning enables rapid identification of potent DDR1 kinase inhibitors. *Nat Biotechnol, 37*(9), 1038-1040. https://doi.org/ 10.1038/s41587-019-0224-x.

Amit Gangwal, A. A. (2024). Generative artificial intelligence in drug discovery: basic framework, recent advances, challenges, and opportunities. *Front Pharmacol, 15*. https://doi.org/10.3389/fphar.2024.1331062.

Atul Sharma, S. T. (2020). Severe acute respiratory syndrome coronavirus-2 (SARS-CoV-2): a global pandemic and treatment strategies. *International Journal of Antimicrobial Agents, 56*(2). https://doi.org/https://doi.org/10.1016/j.ijantimicag.2020.106054.

Burki, T. (2020). A new paradigm for drug development. *Lancet Digit Health, 2*(5), e226–e227. https://doi.org/10.1016/S2589-7500(20)30088-1.

Dalwinder Singh, B. S. (2020). Investigating the impact of data normalization on classification performance. *Applied Soft Computing, 97*. https://doi.org/https:// doi.org/10.1016/j.asoc.2019.105524.

Deep Q-Learning. (2023, Jan 23). Retrieved from Geeks for Geeks: https://www. geeksforgeeks.org/deep-q-learning/.

Greig, R. C. (2017). Drug discovery and development: Role of basic biological research. *Alzheimers Dement (N Y), 3*(4), 651–657.

Hannes H. Loeffler, J. H. (2024). Reinvent 4: Modern AI–driven generative molecule design. *Journal of Cheminformatics, 16*.

Hodos, R. (2021, Jul 13). *AI In Drug Discovery.* Retrieved from BenevolentAI: https://www.benevolent.com/news-and-media/blog-and-videos/ai-drug-discovery/.

Hwanhee Kim, S. K. (2022). Predicting chemical structure using reinforcement learning with a stack-augmented conditional variational autoencoder. *J Cheminform, 14*. https://doi.org/10.1186/s13321-022-00666-9.

Izhar Wallach, M. D. (2015, Oct 10). *AtomNet: A Deep Convolutional Neural Network for Bioactivity Prediction in Structure-based Drug Discovery.* Retrieved from Cornell University: https://arxiv.org/abs/1510.02855.

Khader Hamdia, X. Z. (2021). An efficient optimization approach for designing machine learning models based on genetic algorithm. *Neural Computing and Applications, 33*, 1923–1933. https://doi.org/10.1007/s00521-020-05035-x.

LeCun, Y. (2019). 1.1 Deep Learning Hardware: Past, Present, and Future. *2019 IEEE International Solid- State Circuits Conference - (ISSCC).* San Francisco, CA, USA: IEEE Xplore. https://doi.org/10.1109/ISSCC.2019.8662396.

Manan Goel, S. R. (2021). MoleGuLAR: Molecule Generation Using Reinforcement Learning with Alternating Rewards. *Journal of Chemical Information and Modeling, 61*(12), 5815-5826. https://doi.org/10.1021/acs.jcim.1c01341.

Marcelo C. R. Melo, J. R. (2021). Accelerating antibiotic discovery through artificial intelligence. *Commun Biol.* https://doi.org/10.1038/s42003-021-02586-0.

Chapter 15

Dynamic Drug Discovery: Programming with Real-Time Example Workflows

Kolla Bhanu Prakash*
Department of Computer Science and Engineering,
Koneru Lakshmaiah Education Foundation Deemed to be University,
Green Fields, Vaddeswaram Guntur, Andhra Pradesh, India

Message Passing Neural Networks for Molecule Property Prediction

Message Passing Neural Networks (MPNNs) are a class of neural networks particularly well-suited for tasks involving graph-structured data, like molecule property prediction. In this context, each molecule is represented as a graph where atoms are nodes and bonds are edges. MPNNs can learn to predict properties of molecules by iteratively passing messages between neighboring nodes (atoms) in the graph.

```
using Pkg
    Pkg.add(["Flux", "Graphs", "MetaGraphs", "DataFrames"])

using DataFrames
    # Example: Load your dataset
    df = DataFrame(CSV.File("your_molecule_data.csv"))
```

* Corresponding Author's Email: drkbp1981@gmail.com

In: Harnessing AI for Drug Discovery
Editor: Kolla Bhanu Prakash
ISBN: 979-8-89530-540-9

```
using Graphs
    # Example: Create a graph for a molecule
    g = SimpleGraph(5) # Assume 5 atoms in the molecule
    add_edge!(g, 1, 2) # Add bonds between atoms
    add_edge!(g, 2, 3)

using Flux
    struct MPNN
    layers::Vector{Any}
    end

    function (mpnn::MPNN)(graph, node_features)
     for layer in mpnn.layers
     node_features = layer(graph, node_features)
     end
     return readout(node_features)
    end

    function message_passing_layer(graph, node_features)
     new_node_features = node_features * some_weight_matrix
     for (i, j) in edges(graph)
     # Aggregate messages from neighbors
     new_node_features[j] += some_function(node_features[i])
     end
     return new_node_features
    end

    function readout(node_features)
     return sum(node_features, dims=1)
    # Example: sum node features for the final prediction
    end

    mpnn = MPNN([message_passing_layer, message_passing_layer,
readout])

    loss(y_pred, y_true) = Flux.mse(y_pred, y_true)
    opt = Flux.ADAM(0.001)

    for epoch in 1:100
```

```
 Flux.train!(loss, params(mpnn), data, opt)
end
```

Use metrics like RMSE or MAE to evaluate the model's performance on a validation set. Once trained, use the model to predict properties of new molecules by feeding in their graph representations. Experiment with different network architectures, message-passing steps, and learning rates. Explore augmenting your data with more molecular features or different graph representations. This approach should give you a good starting point for implementing MPNNs for molecular property prediction using Julia.

Deep Graph Learning for Drug Discovery

This project implements a Message Passing Neural Network (MPNN) in Julia to predict molecular properties for drug discovery. The MPNN operates on graph-structured data where molecules are represented as graphs with atoms as nodes and bonds as edges.

```
using Pkg
    Pkg.add([“Flux”, “Graphs”, “MetaGraphs”, “DataFrames”, “CSV”,
“Random”, “CUDA”])

├── data/
│  └── molecules.csv # Example dataset
├── src/
│  ├── data_loader.jl # Code for loading and preprocessing data
│  ├── model.jl # MPNN model implementation
│  ├── train.jl # Training loop
│  └── evaluate.jl # Model evaluation
├── run.jl # Main script to run the project
└── README.md # Project documentation

module DataLoader
using CSV, DataFrames, Graphs, Random
export load_data, process_molecule

function load_data(filepath::String)
 df = DataFrame(CSV.File(filepath))
```

```
 return df
end

function process_molecule(row)
 num_atoms = row[:num_atoms]
 bonds = row[:bonds] # Assume this is a list of tuples representing bonds

 g = SimpleGraph(num_atoms)
 for bond in bonds
 add_edge!(g, bond[1], bond[2])
 end

 node_features = rand(Float32, num_atoms, 10) # Example: random node
features
 return g, node_features
end
end
module Model
using Flux, Graphs
export MPNN, message_passing_layer, readout_layer
struct MPNN
 layers::Vector{Any}
end
function (mpnn::MPNN)(graph, node_features)
 for layer in mpnn.layers
 node_features = layer(graph, node_features)
 end
 return readout_layer(node_features)
end
function message_passing_layer(graph, node_features)
 new_node_features = node_features
 for (i, j) in edges(graph)
 # Aggregate messages from neighbors
 new_node_features[j, :] += node_features[i, :] * rand(Float32,
size(node_features, 2), size(node_features, 2))
 end
 return new_node_features
end
```

```
function readout_layer(node_features)
 return sum(node_features, dims=1)
end

mpnn = MPNN([message_passing_layer, message_passing_layer])
end
module Train
using Flux, DataLoader, Model, Random
export train_model
function train_model(df, epochs=100, lr=0.001)
 opt = Flux.ADAM(lr)

 for epoch in 1:epochs
 total_loss = 0.0
 for row in eachrow(df)
 g, node_features = DataLoader.process_molecule(row)
 y_true = row[:target] # Assume the target property is stored in :target

 y_pred = Model.mpnn(g, node_features)
 loss = Flux.mse(y_pred, y_true)
 total_loss += loss
 Flux.train!(loss, params(Model.mpnn), [(g, node_features)], opt)
 end
 println("Epoch $epoch, Loss: $total_loss")
 end
end
end
module Evaluate
using Flux, DataLoader, Model
export evaluate_model
function evaluate_model(df)
 total_loss = 0.0

 for row in eachrow(df)
 g, node_features = DataLoader.process_molecule(row)
 y_true = row[:target]

 y_pred = Model.mpnn(g, node_features)
 loss = Flux.mse(y_pred, y_true)
```

```
 total_loss += loss
 end
 println("Evaluation Loss: $total_loss")
end
end
using CSV
using DataLoader
using Train
using Evaluate
# Load data
train_df = DataLoader.load_data("data/molecules_train.csv")
test_df = DataLoader.load_data("data/molecules_test.csv")

# Train the model
Train.train_model(train_df, epochs=100, lr=0.001)

# Evaluate the model
Evaluate.evaluate_model(test_df)
```

Building E(3)-Equivariant Interatomic Potentials Using NequIP

NequIP is a deep learning framework designed to build interatomic potentials that are equivariant to 3D rotations and translations. This is particularly useful for molecular dynamics simulations, where it's essential to capture the symmetries of physical systems.

Make sure you have the required dependencies installed. You can set up a virtual environment and install NequIP via pip.

```
python3 -m venv nequip-env
source nequip-env/bin/activate
pip install torch nequip
```

NequIP requires a dataset that contains atomic positions and corresponding energy and force labels. If you don't have one, you can use the example datasets provided in the NequIP repository or generate your own using quantum chemistry software like VASP or Quantum ESPRESSO.

Create a YAML configuration file specifying the model architecture, training parameters, and dataset location.

```
# config.yaml
dataset: "path/to/your/dataset"
model:
 r_max: 5.0
 n_radial_basis: 8
 n_hidden: 64
 n_layers: 3
 nonlinearity: swish
optimizer:
 lr: 0.001
 weight_decay: 1e-5
training:
 batch_size: 32
 n_epochs: 100
```

Use the nequip-train command to train the model using your dataset and configuration file. This will output a trained model file that you can use for inference.

```
nequip-train config.yaml
```

After training, you can evaluate the model on a test set or use it to predict the properties of new atomic configurations.

```
from nequip.data import AtomicData
from nequip.nn import NequIPModel
import torch

# Load the trained model
model = torch.load("path/to/trained_model.pth")

# Example: Predict properties of a new atomic configuration
atomic_data = AtomicData.from_pos(positions=new_positions, species=new_species)
prediction = model(atomic_data)
print(prediction["energy"], prediction["forces"])
```

You can visualize the predicted energies and forces and compare them with the reference data to ensure the model's performance. NequIP is highly customizable. You can tweak the model's architecture, training process, and dataset preprocessing to suit your specific project needs. Ensure your dataset is well-prepared and contains sufficient examples for the model to learn accurate interatomic potentials.

Creating Interaction Fingerprints (IFPs) for Protein-Ligand Complexes

Creating Interaction Fingerprints (IFPs) for protein-ligand complexes using Julia involves a few steps, including loading the protein-ligand complex data, calculating interactions, and then generating the fingerprints.

First, ensure that you have Julia installed, along with some necessary packages. You might need to use packages like BioStructures.jl for handling PDB files and DataFrames.jl for managing the fingerprint data.

```
using Pkg
Pkg.add([“BioStructures”, “DataFrames”, “Distances”])
```

Start by loading the protein-ligand complex from a PDB file.

```
using BioStructures

# Load the protein-ligand complex from a PDB file
structure = read(“path/to/protein_ligand_complex.pdb”, PDB)
```

You need to separate the ligand and protein atoms for interaction calculations.

```
function extract_atoms_by_residue(structure, residue_name)
 return filter(atom -> atom.resname == residue_name, collect(atoms(structure)))
end

ligand_atoms = extract_atoms_by_residue(structure, “LIG”)
# Assuming LIG is the ligand residue name
protein_atoms = filter(atom -> atom.resname != “LIG”, collect(atoms(structure)))
```

Common interactions include hydrogen bonds, hydrophobic contacts, and electrostatic interactions. For simplicity, let's focus on distance-based interactions.

```
using Distances

function calculate_interactions(ligand_atoms, protein_atoms, cutoff_distance=5.0)
 interactions = []
 for lig_atom in ligand_atoms
 for prot_atom in protein_atoms
 dist = pairwise(Euclidean(), [lig_atom.xyz], [prot_atom.xyz])[1]
 if dist < cutoff_distance
 push!(interactions, (lig_atom.name, prot_atom.name, dist))
 end
 end
 end
 return interactions
end

interactions = calculate_interactions(ligand_atoms, protein_atoms)
```

You can represent the interactions as a binary vector where each position corresponds to a specific type of interaction.

```
using DataFrames
# Define interaction types (e.g., hydrophobic, hydrogen bond, etc.)
interaction_types = ["Hbond", "Hydrophobic", "Electrostatic"]
interaction_fingerprint = zeros(Int, length(interaction_types))

# Example: Simple rule-based classification
for (lig_atom, prot_atom, dist) in interactions
 if dist < 3.5
 interaction_fingerprint[1] = 1 # Example: Hydrogen bond
 elseif 3.5 <= dist < 5.0
 interaction_fingerprint[2] = 1 # Example: Hydrophobic interaction
 end
 # Add more rules as needed
end
```

```
df=DataFrame(interaction_types=interaction_types,interaction_fingerprint=interaction_fingerprint)
println(df)
```

Once you have the fingerprints, you can analyze or visualize them depending on your project requirements.

```
using Plots

bar(df.interaction_types, df.interaction_fingerprint, legend=false, xlabel="Interaction Type", ylabel="Presence")
```

Combining Generative AI, Molecular Docking, and Molecular Dynamics (MD) Simulations for the Identification of Novel Inhibitors

Data Preparation: Obtain a dataset of known EthR inhibitors and their structures.

Generative AI Model: Train a generative model to design novel compounds similar to known inhibitors.

Molecular Docking: Dock the generated molecules to the EthR protein to identify potential binders.

Molecular Dynamics Simulations: Simulate the docked complexes to evaluate their stability.

Analysis: Analyze the results to identify the most promising candidates.

You can use SMILES strings to represent the molecules.

```
# Example SMILES for known EthR inhibitors
known_inhibitors=["CC1=CC=C(C=C1)C2=NN=C(S2)N","CCOC(=O)C1=CC=C(C=C1)N2C=NN=C2C3=CC=CC=C3"]
```

Training the generative model to design novel compounds similar to known inhibitors is given below.

```
import torch
from rdkit import Chem
from rdkit.Chem import AllChem
from torch import nn
import random

# Example Variational Autoencoder (VAE) for molecule generation

class MoleculeVAE(nn.Module):
 def __init__(self):
 super(MoleculeVAE, self).__init__()
 self.encoder = nn.Sequential(
 nn.Linear(256, 128),
 nn.ReLU(),
 nn.Linear(128, 64)
 )
 self.decoder = nn.Sequential(
 nn.Linear(64, 128),
 nn.ReLU(),
 nn.Linear(128, 256),
 nn.Sigmoid()
 )

 def forward(self, x):
 z = self.encoder(x)
 return self.decoder(z)

def smiles_to_fingerprint(smiles):
 mol = Chem.MolFromSmiles(smiles)
 fp = AllChem.GetMorganFingerprintAsBitVect(mol, 2, nBits=256)
 return torch.Tensor(fp)

# Train VAE with known inhibitors
vae = MoleculeVAE()
optimizer = torch.optim.Adam(vae.parameters(), lr=0.001)
loss_fn = nn.MSELoss()
```

```
for epoch in range(100):
 for smiles in known_inhibitors:
 x = smiles_to_fingerprint(smiles)
 x_pred = vae(x)
 loss = loss_fn(x_pred, x)
 optimizer.zero_grad()
 loss.backward()
 optimizer.step()

# Generate new molecules
new_molecules = []
for _ in range(10):
 z = torch.randn(64)
 new_fp = vae.decoder(z)
 new_molecules.append(new_fp)

# Convert fingerprints back to SMILES (RDKit example)
def fingerprint_to_smiles(fp):
 mol = AllChem.RDKFingerprintToMol(fp)
 return Chem.MolToSmiles(mol)

generated_smiles = [fingerprint_to_smiles(fp) for fp in new_molecules]
```

We can use AutoDock Vina, which is typically called via command-line or Python bindings.

```
import os

# Prepare docking with AutoDock Vina
for smiles in generated_smiles:
 mol = Chem.MolFromSmiles(smiles)
 Chem.MolToPDBFile(mol, "generated_ligand.pdb")
 os.system("obabel generated_ligand.pdb -O generated_ligand.pdbqt")
 os.system("vina --receptor EthR.pdbqt --ligand generated_ligand.pdbqt -
-config config.txt --out output.pdbqt --log log.txt")
```

Using a tool like GROMACS, you can set up MD simulations for the docked complexes.

```
# GROMACS commands (assuming you have the necessary setup files)
gmx pdb2gmx -f output.pdbqt -o complex.gro -water spce
gmx editconf -f complex.gro -o newbox.gro -c -d 1.0 -bt cubic
gmx solvate -cp newbox.gro -cs spc216.gro -o solv.gro -p topol.top
gmx grompp -f ions.mdp -c solv.gro -p topol.top -o ions.tpr
gmx genion -s ions.tpr -o solv_ions.gro -p topol.top -pname NA -nname
CL -neutral
gmx grompp -f minim.mdp -c solv_ions.gro -p topol.top -o em.tpr
gmx mdrun -v -deffnm em

# Run NVT and NPT equilibration
gmx grompp -f nvt.mdp -c em.gro -r em.gro -p topol.top -o nvt.tpr
gmx mdrun -deffnm nvt
gmx grompp -f npt.mdp -c nvt.gro -r nvt.gro -p topol.top -o npt.tpr
gmx mdrun -deffnm npt

# Run production MD
gmx grompp -f md.mdp -c npt.gro -p topol.top -o md.tpr
gmx mdrun -deffnm md
```

You can analyze the MD trajectories to assess the stability and binding affinity of the generated compounds.

```
import MDAnalysis as mda
u = mda.Universe("topol.top", "md.trr")
protein = u.select_atoms("protein")
ligand = u.select_atoms("resname LIG")
rmsd = []
for ts in u.trajectory:
 rmsd.append(protein.rmsd(protein))
import matplotlib.pyplot as plt
plt.plot(rmsd)
plt.xlabel("Time (ps)")
plt.ylabel("RMSD (Å)")
plt.title("Protein-Ligand Complex Stability")
plt.show()
```

This end-to-end example combines Julia, Python, and GROMACS for generative design, docking, and molecular dynamics of EthR inhibitors targeting Mycobacterium tuberculosis.

About the Editor

Dr. Kolla Bhanu Prakash is a Professor and Associate Dean [R&D] at K.L. Deemed to be University. He is an Adjunct Professor at Taylors University, Malaysia. He has published 150+ research papers in international and national journals and conferences. He published Scopus and SCI publications 100 with an H-index of 23 and a total citation of 1600. He is a reviewer for several international journals, including IEEE, Elsevier, Springer, Wiley, Taylor, and Francis, and he has served as a reviewer, keynote speaker, and TPC member for several international conferences. His research interests include deep learning, data science, and quantum computing. He received the Best Researcher Award 4 times. He authored three and edited 15 books with internationally reputed publishers like IEEE, Elsevier, Springer, River, CRC, Degryuter, and Wiley. He also published 15 patents, two patent grants, and 1 Copyright. He is guiding 3 PhD scholars. Under his guidance, 3 PhD's were awarded. Dr. K. Bhanu Prakash is an IEEE senior member, fellow ISRD, and treasurer of the ACM Amaravathi Chapter in India, LMISTE, MIAENG, and SMIRED. He is the Series Editor of "Next Generation Computing & Communication Engineering" at Wiley Publishers. He is a series editor - "Industry 5.0: Artificial Intelligence, Cyber-Physical Systems, Mechatronics, and Smart Grids" – CRC publishers. He is a NAAC Accessor. He is listed in the top 2% of cited researchers worldwide by Stanford University and Elsevier for 2021. He received a grant of 27 lakhs from Meity for establishing a Quantum Computing Research lab [QCAL] at KLEF and a grant from DST SERB worth five lakhs for organizing a workshop and conference in Quantum Computing. He delivered many keynote speeches and expert talks worldwide on quantum computing, machine learning, and deep learning.

Index

E

G

H

I

L

M